S0-BWX-547

Mining in the New World

Mining in the New World

by

Carlos Prieto

Consulting Editor
Marvin D. Bernstein

Sponsored by The Spanish Institute, Inc.

McGraw-Hill Book Company

New York St. Louis San Francisco
Dusseldorf Johannesburg Kuala Lumpur London
Mexico City Montreal New Delhi Panama
Rio de Janeiro Singapore Sydney Toronto

Library of Congress Cataloging in Publication Data

Prieto, Carlos, 1898–
Mining in the New World.

Translation of La minería en el Nuevo Mundo.
"Sponsored by the Spanish Institute."
Bibliography: p.
1. Mines and mineral resources—Latin America—
History. I. Spanish Institute. II. Title.
TN27.5.P713 338.2'098 72-4245
ISBN 0-07-050862-3

123456789

To the illustrious body of Mexican mining engineers
of the past and of the present

Contents

Foreword

The Spanish Institute is a private organization founded in 1954 to promote interest in Hispanic culture and understanding between the people of the Old and the New Worlds. Among its original directors were an architect, an anthropologist, an art historian, a book collector, a sculptor, an opera singer, a surgeon, a lawyer, a banker, a diplomat, and a writer. Its New York headquarters houses a library, a book store, and an art gallery, and as part of its ever-broadening educational program, the Institute offers courses in Spanish, a series of concerts and films, and a wide variety of lectures. The Institute also supports medical research fellowships and archeological excavations in Spain. With such broad objectives of its own, it seems especially fitting that the Spanish Institute should sponsor the publication of a book by a man of such varied interests and talents as the distinguished Mexican of Spanish birth, Carlos Prieto.

Carlos Prieto was born in 1898 at Oviedo in northern Spain. Possessing a keen mind and exceptional powers of concentration, he developed his faculties and tastes in many directions at an early age. Although he might well have specialized in economics, science, agriculture, letters, or music, Sr. Prieto finally decided to study jurisprudence, the science of the rights and responsibilities of man. After receiving a remarkable secondary education, he entered the University of Oviedo where he obtained the degree of licenciado in law. In 1923 Licenciado Prieto moved to Mexico and soon was qualified by the Universidad Autónoma Nacional in Mexico City to practice law in his adopted land. Not long afterward, he became head of the legal division of the Compañia Fundidora de Fierro y Acero de Monterrey, S.A. (Monterrey Iron and

Steel Co., Inc.). In 1931 he was named a member of the board, and in 1945 he became Chairman, a post he still holds.

Licenciado Prieto, who was one of the founders of the Instituto Latinoamericano del Fierro y del Acero and its president for five years, is deeply concerned with mining and has been extraordinarily active in today's business. But he also has deep feelings for the pageant of the past, and has had an unrivaled opportunity to travel widely, think profoundly, and explore the role of mining in the development of man's environment. As Pedro Lain Entralgo wrote in his preface to the Spanish edition of this work:

> There is no true Spaniard who has not, after his fashion, experienced in his soul the reality and history of Spanish America; there is no true American who, having been born in one of the countries of that America, has not experienced for himself the works of the Spaniards who formed the history and reality of his country in such great measure. There can be very few who have matched Carlos Prieto in the fulfillment of this double requisite. He arrived in Mexico as a Spaniard, nearly fifty years ago; with a Spaniard's love he began to live the exciting reality of the great country of Mexico; and today, as a citizen of that country and indeed of all Spanish America, he presents to us in a most original and well-documented form, one of the most important and thought-provoking aspects of American history—the eminent role of mining in the genesis of the nations which today constitute the New World.

Carlos Prieto's thesis, like that of Humboldt in this matter, defends the view that mining, for all its hardships and injustices, has contributed enormously to the civilizing of the world. A few years after H. G. Wells published his *Outline of History,* another Huxley student, Thomas A. Rickard, brought out a two-volume work entitled *Man and Metals: A History of Mining in Relation to the Development of Civilization.* Rickard felt that his fellow student had failed to "pay proper regard to the part mining had played." In discussing the New World, Rickard rightly observed that the Spaniards did in Peru and Mexico what the Carthaginians and Romans had done in Spain: they compelled the natives, as well as their imported slaves, to work the mines for them.

When mentioning the Spanish and Portuguese search for precious gems and metals, non-Iberian historians and reformers have tended to stress the cruel and inhuman treatment meted out to the workers—Indian, Negro, mestizo, and white—and to minimize the importance and effectiveness not only of the laws enacted to protect the workers, but also of the widespread, unquestionable, and lasting benefits of mining. Reflecting in a recent article on the "Black Legend" and the treatment of workers by their Spanish and Portuguese masters, Lewis Hanke calls for a moratorium on generalizations and criticizes "pragmatic" writers who put on twentieth-century spectacles when analyzing Iberian colonial practices. As one who has devoted years to the study of Las Casas, a champion leading the struggle for justice, as well as to the study of the history of the "Imperial City" of Potosí, Hanke is acquainted with both sides of the argument and his advice should be heeded.

Until recently, there have been few serious modern studies on the history of ore extraction, and an American mining engineer, Herbert Hoover, was one of the first to become interested in its technical evolution. In 1912 Mr. Hoover and his wife brought out a facsimile of Georgius Agricola's *De re metallica* (originally written in Chemnitz in 1556) with an exemplary translation and elaborate notes. This beautifully illustrated treatise, which describes numerous mining and refining techniques, was written before the German mineralogist could have become acquainted with the concept of reducing silver by the "patio process," which was developed in Mexico in 1554. Mr. Hoover notes this fact and comments on Fray Alvaro Alonso Barba's *El arte de los metales* (Madrid, 1640): "Among the methods of silver amalgamation described by Barba is one which, upon later 'discovery' at Virginia City (in the 1860's) is now known as the 'Washoe Process.' " As President-elect of the United States, Mr. Hoover visited Latin America in 1928 and outlined plans for the greater mutual understanding between English- and Spanish- and Portuguese-speaking people of the hemisphere. Indeed, Hoover's views on cooperation preceded those of Franklin D. Roosevelt's Good Neighbor Policy.

While Carlos Prieto has not translated early treatises on metallurgy, his company has issued facsimile editions of Nicolas Monardes' *Diálogo del hierro y sus grandezas* (Seville, 1574) as well as the 1770 edition of Barba's book. He has also been a long-time advocate of closer inter-American ties. In fact, by being a patron of the Spanish Institute in New York and by writing this book, he is in one more way furthering international understanding, as Mr. Hoover did.

An admirer of the Spanish version of Lic. Prieto's study, the eminent Hispanist and cultural historian Irving A. Leonard (until recently Domingo F. Sarmiento Professor of Spanish American History and Literature at the University of Michigan), has written of its "succinct" and "illuminating" presentation. His concluding observations should be noted here:

> The productive economic activities of colonial Hispanic and Portuguese America were: agriculture and pastoral pursuits, mining and crafts, and their importance was in that order. But clearly it was the liquid wealth, mainly silver and gold, provided by the mines particularly in Mexico, Peru and Brazil, that caused other activities to flourish. The mining camps required farms to feed workers, cattle ranches for hides and tallow, mules, horses and improved roads for transportation, while the silver and gold of the mines and streams stimulated trade, crafts, urbanization, the growth of an aristocratic society with an incipient middle class, the building of ornate ecclesiastical and municipal structures as well as private residences of Baroque ostentation, and a cultivation of the arts and enjoyment of luxury that exceed anything that the neighboring English colonies knew, especially in the seventeenth century.

In closing, a few words more should be said about the author, who is not only a mining expert and a lawyer but also an agriculturist, a botanist, a zoologist, a bibliophile, and an authority on music. Because of his efforts the economy of the valley of the Tecolutla River in the state of Veracruz has been transformed, and today there are over 5 million citrus trees in the region. His concern for natural history has resulted in the establishment of a mollusk museum and a tropical botanical garden in Tecolutla and the sponsorship of numerous studies by zoologists and botanists. A number of rare zoological specimens bear his name. Carlos Prieto's rich collection of books on Don Quijote and Cervantes has already been given to the Technological Institute of Monterrey, and he has instituted an annual prize for the best book on the Spanish language. All in all, Sr. Prieto is a man of multiple talents. His broad grasp of many fields of knowledge has contributed much to the enrichment of this study.

Carleton Sprague Smith
The Spanish Institute, Inc.

Consulting Editor's Note

It is indeed an honor to be asked to act as Consulting Editor for this English version of Licenciado Carlos Prieto's study, *La minería en el Nuevo Mundo.* While I do not subscribe to all of Lic. Prieto's theses and conclusions, I believe that he has presented a strong case for his point of view which deserves to be read and studied by a large audience of students and readers. It is truly unfortunate that biases directed against the Iberian peoples and their labors in the New World still exist, that there are people who would downgrade their work by condemning the Hispanic colonizers on the basis of a "Black Legend" of 350 years of unrelieved cruelty, plunder, and exploitation. To these condemners there has been added recently the New Left's criticism of all colonial powers. They hold that it is wrong to impose alien cultures and controls upon native civilizations since those civilizations had a right to exist uninfluenced by outside social and spiritual values and technologies. Such simplistic attitudes have to be refuted if we are to have a picture of history closer to the truth.

By approaching the study and interpretation of Hispanic American history from the standpoint of his special knowledge of and interest in mining, Lic. Prieto has been able to add an often-neglected dimension and to point out frequently overlooked relationships between the history and culture of Hispanic America and Iberian economic and technological activities. The result is a truly refreshing viewpoint.

I should like to thank Dr. Carleton Sprague Smith for his comments and suggestions, particularly in connection with the arts and the material on Brazil. The chapter on legal edicts and the bibliography are mostly my work, with helpful suggestions from Lic. Prieto.

MARVIN D. BERNSTEIN
State University of New York at Buffalo

Introduction

During the last three months of 1967, the College of Engineering of the National University of Mexico organized a number of events to celebrate the one hundred seventy-fifth anniversary of the founding in 1792 of the Royal College of Mining in Mexico City and the first centenary of its transformation into a school of engineering. Among these events was a cycle of lectures having to do with the school and with Mexican engineering in general. I was invited to participate by delivering a lecture. The subject of my talk, delivered on October 31, 1967, was "The Importance of Mining in the Discovery of the New World and the Formation of Ibero-American Nations." The lecture was a summary of my historical thinking as a man who has experienced Spanish American life very intensively in devoted service to Mexico. In that role I have traveled through all the Spanish-speaking countries of America, as well as Brazil, and have thought deeply about the meaning of the accomplishments of Spain and Portugal on this continent. My research in preparation for that address constitutes the basis from which this book grew.

At the request of the organizers of those commemorative ceremonies and of many who heard the lecture, I decided to publish my commentary, but in a greatly augmented form. In this book, I have been able to include both the material which a lecturer is obliged to pass over in order to remain within his allotted time and the numerous complementary observations and bibliographical notes which may serve as a guide to the reader interested in pursuing the subject more deeply.

I must mention here my gratitude to a number of persons who read the original manuscript and helped me with useful observations which improved the text: Don Modesto Bargalló, professor at the National School of Biological Sciences of the National Polytechnic Institute; Dr. Arturo Arnaiz y Freg, member of the Academy of History; Professor Florentino M. Törner, critic and essayist; Don José Iturriaga, economist, historian, and journalist; and Don León Salinas and Don Gustavo P. Serrano, distinguished engineers and former Ministers in our government. In preparing this greatly expanded English version of the original Spanish edition of my study, I am deeply indebted to the Consulting Editor, Dr. Marvin Bernstein of the State University of New York at Buffalo, and to Dr. Carleton Sprague Smith, Vice Chairman of the Spanish Institute, for their generous assistance and many perceptive suggestions for augmenting and strengthening my original thesis.

Dr. Smith, a veteran Hispanist and Latin American authority, studied at the Centro de Estudios Historicos in Madrid, and his investigations in Spanish and Portuguese archives and centers in countries south of the Rio Grande early made him familiar with the thesis of this study. Having lived four years in Brazil, he was particularly qualified to furnish significant data for the sections dealing with Lusitanian America. Again, as professor of history at New York University for over two decades, he helped remodel the book for English-speaking readers. Finally, his special knowledge of music and fine arts in their relation to mining give those chapters added meaning. His overall interest and enthusiasm have been heart-warming to the author.

Marvin Bernstein, a specialist in mining history, obtained his Ph.D. from the University of Texas and has been active in the field for many years. He was a visiting professor at the Central University of Ecuador, Quito, during 1957–1958 and at the Instituto Tecnológico of Monterrey, Mexico, in 1966. A member of the Pan American Institute of Geography and History, he is the author of *The Mexican Mining Industry, 1890–1950* (1965). In 1970 he was appointed professor of history at the University of Buffalo. His familiarity with mining and its ramifications make him an ideal technical editor for the English edition of this work.

For the reader with a greater curiosity concerning the history of Hispanic American mining, I should like to point out the work of the Compañía Fundidora de Fierro y Acero de Monterrey, S.A., with which I have been associated since 1923 and of which I am now President, in making available reprints of classic works as well as scholarly studies in the field. Among others, the company has printed at its expense: (1) in 1925, a facsimile edition of Alvaro Alonso Barba's *El arte de los metales,* 6th Spanish ed. (Madrid, 1770), whose plates were presented to the Escuela Superior de Minas in Madrid, which used them to make a new printing in 1932; (2) Modesto Bargalló, *La minería y la metalurgia en la America Española durante la época colonial* (Mexico City: Fondo de Cultura Económica, 1955), on the fourth centenary of Bartolomé de Medina's discovery of the amalgamation process; (3) *Diálogo del hierro y sus grandezas* by Doctor Monardes, a physician of Seville, in a facsimile of the 1574 edition (Seville), and a version in modern Spanish of the same work (Mexico City, 1961); (4) Modesto Bargalló, *Las ferrerías de los primeros veinticinco años del Mexico independiente* (Mexico City, 1965), and also his *Andrés Manuel del Río y su obra científica* (Mexico City, 1966), on the second centenary of his birth; and finally, the highly important work of Modesto Bargalló, *La amalgamación de los minerales de plata (Amalgamation of Silver Ores)* (Mexico City, 1970) to commemorate the fourth centenary of the birth of Alvares Alonso Barba.

Carlos Prieto

Mining in the New World

I

Thesis of This Study

A short time ago I was in the famous "Noble and Most Faithful Imperial City of Potosí" at the invitation of Don Armando Alba, an outstanding Bolivian and President of the Geographical and Historical Society of Potosí. On that visit I was moved to comment that if the mines of the Cerro Rico of Potosí had been the base of the Spanish Empire, then Spanish and Ibero-American mining had also created twenty nations which, through their personality and culture, form part of what we are accustomed to call the Western World. My statement was not merely a polite phrase spoken in that impressive setting high among the Andean ranges. It was a conclusion reached after long studies, observations, and experiences personally garnered in all the Iberian countries of the continent concerning the decisive influence of mining on the formation of the Ibero-American nationalities. From these thoughts this study grew.

In the course of this discussion I shall comment on the incredibly short period of time in which the discovery and exploration of the American continent was carried out. The rapidity of this accomplishment can be explained primarily by the influence of reports of the abundance of precious metals. I shall then have something to say about the first miners who arrived in the new lands and the first mines opened for exploitation. This discussion will take us on an imaginary journey through most of the countries of Spanish and Portuguese America, from the Mexican Sierras Madres to the southernmost tip of the Andes, in the territory of Chile and the island of Tierra del Fuego, where the mountains dissolve into myriad lakes, fiords, and islands. I shall mention the impetus that mining gave to the development of agriculture and trade, to the building of roads and ports, to the formation of cities, and to the establishment of educational institutions. I shall refer to the techniques applied to metallurgy, as well as to the mining ordinances and legal institutions that governed and administered this fundamental activity in the New World. And I shall end by emphasizing an extraordinarily important phenomenon: the participation in the Emancipation Movement by mining engineers, many of whom gave their lives for their ideals.

The central thesis of this study is that mining was the creator of the peoples and nations of Ibero-America as they exist today. The discovery of the New World and the rapid exploration and occupation of an *entire* continent in less than sixty years was due to the fact that, as early as Columbus's first voyage, proof was found of the presence of gold. This discovery gave impetus to the search for and exploitation of deposits of precious metals. As a result, peoples and nations which now constitute a very important part of the Western World were formed in these new lands, peoples and nations with personalities of their own and with a culture containing both Hellenic and Christian elements.

Around this theme I have sketched a number of subsidiary theses in support of my assertion. As a sort of index, I have listed some of them:

1. The search for gold and other precious metals has always been a human occupation. It has manifested itself in literature, myth, and human life throughout the history of mankind, from the legend of Jason and the Argonauts to the most recent disturbances in the international monetary field.
2. Though at the outset there were excesses in the appropriation of treasures, and later in the exploitation of mines—there is no reason to deny that there was plunder—we must keep in mind the attitudes of the period toward war and conquest and the special position of discoverers. But let us not forget that there also were legal regulations for protection of the aborigines drawn up by the Crown and the Council of the Indies. There were hosts of missionaries who eased the settlers' abuses. And there was complete freedom in both America and Spain to criticize severely evasions of the law as well as ill-

treatment of the natives. (The most outstanding critic was Fray Bartolomé de Las Casas.)

3. Discovery and exploitation of mines was not an easy task. It required a number of unusual qualities and skills which culminated in the invention and application of new metallurgical techniques that made it possible to obtain fabulous amounts of gold and silver. These techniques merited the praise of well-known European technicians as late as the beginning of the nineteenth century.
4. The large shipments of gold and silver from the New World between the sixteenth and eighteenth centuries had a tremendous influence on the development of Europe.
5. Not all the precious metal was sent to the Old World, however. Much remained in the new territories and formed the basis for the development of the Ibero-American peoples. It is worth noting that the Viceroyalty of Peru was responsible for shipping the so-called *situados* (bars of metal or minted coins) to Chile and Río de la Plata and that over a period of three centuries the Viceroyalty of New Spain gave financial aid to Panama, Colombia, Venezuela, Florida, Cuba, Puerto Rico, Jamaica, and even the far-away Philippine Islands.
6. Despite frequent contentions to the contrary, mining neither excluded nor aborted the development of other economic activities such as agriculture, trade, and industry. Rather, the mining industry constituted a stimulus and support for these other activities, and within the technical limitations of the time, it made possible a network of roads from Upper California and Texas to Tierra del Fuego and over the mountains from the Pacific to the Atlantic.
7. Spanish colonization bore little resemblance to "colonialism" as the word is used in modern times. The modern version is characterized by exploitation of a country's resources at the expense of the natives, with whom there is little mixing of either

blood or culture. It exists solely for the benefit of the mother country and the colonialists (not settlers) who come with the intention of eventually returning to their own country. In contrast, the Spaniards and Portuguese who left the Peninsula and came to the New World made their homes in the new lands. This philosophy of life was responsible for the construction of large and beautiful cities and for the attention paid to educational institutions. Because of this attitude, the printing press was brought to Mexico as early as 1535, and the Universities of Mexico City and San Marcos at Lima were founded in 1551, thirty years after the Conquest of New Spain and seventeen years after the Conquest of Peru.

8. From the foregoing it can be deduced that almost everything that happened in America after the first generation of conquerors was the accomplishment of "Americans." Except for their responsibility in the establishment of institutions and a political administration, the people of the Iberian Peninsula do not have the right to appropriate the glorious pages of American history. Nor do Ibero-Americans have the right to reject or scorn, as something for which only the Spaniards and Portuguese were to blame, the gray or black pages written during the period of the Viceroyalties.
9. The Independence Movement was an inevitable phenomenon. It did not represent the closing of a parenthesis in the life of indigenous subjugated peoples who at last became free. Rather, emancipation was a natural consequence of the maturing of a society formed by a happy mixture of blood and cultures.
10. The struggle for independence, during which the people's ideas and attitudes were brought out into the open, was a civil war. That fact explains its ferocity, which was neither more nor less than that shown in the French Revolution, the American Civil

War, the Paris Commune of 1870, or the Spanish Civil War of 1936. Hence, also, the fact that when it was over, black memories were left which tarnished the brilliant chapters of the book written by Spain in America over the course of 300 years.

11. My general conclusion, then, is that by some mysterious historical alchemy American gold and silver were transmuted like a philosopher's stone in reverse into intangible essences: into the peoples and nations of Spanish and Portuguese America.

If at the conclusion of this work I have not succeeded in demonstrating my assertions, attribute it to my lack of talent and not to the true meaning of the facts.

II

The Quest for Gold

Before taking up the specific subject of mining in colonial Ibero-America, I should like to examine one often-criticized aspect of the Iberian colonization of the New World: that the Spaniards arrived in those lands, conquered them, and settled them while in the grip of an insatiable desire for riches. As the stereotyped phrase has it, they were mad with a thirst for gold. This assertion, tirelessly repeated by those who wish to discredit the actions of the colonists, implies that until they discovered America and eagerly, skillfully, and successfully sought gold and other precious metals there, all men were indifferent toward or disdainful of those assets, deeming them useless or undesirable. But in truth, always and everywhere, gold and its surrogates have been a goal of human effort. It has been considered necessary and indispensable for trade and for the progress of all peoples. This belief will continue as long as man is man, and it will provide the impetus and motive for his desire for improvement.

It is indeed curious that the strongest criticism of the eager and tireless pursuit of gold and silver deposits in America should have come precisely from those nations that organized piratical enterprises to seize the bullion and coin as booty on the high seas. It was those very nations that rewarded the privateers with titles of nobility that have passed the most severe and acrimonious moral judgment on the men who labored to discover and extract those treasures from the bowels of the Andes and the heart of Minas Gerais.

It should be considered providential that beginning with Columbus's first voyage indications of gold were found in the newly discovered islands and that Columbus was able to take away as a gift for the Catholic Monarchs a few gold nuggets and jewels. If he had not, and if these hopes had not been confirmed later on the mainland, America's fate might have been very different. The countries which were later formed in the New World as a result of the actions of the intrepid men following in the discoverer's footsteps might have been quite unlike those we know today. Because of the activities of such men, these nations possess a culture, a language and literary heritage, a religion, a life-style, a personality, and a spiritual wealth of which they are justly proud.

Even without the lure of precious metals, the Catholic Monarchs would no doubt have continued this enterprise which enlarged their domains and the number of their subjects. Sufficient political, economic, and religious motives existed at a propitious time in the history of the Iberian Peninsula and of Europe. Moreover, having completed their centuries-long task of reconquering their own soil, the Spaniards found in the fabulous discovery of Columbus a powerful stimulus for their imaginations and ample motivation to carry out incredible exploits and adventures. These adventures were possible and real, as distinct from those merely dreamed about by writers and readers of the romances of chivalry on the bare Castillian and Extremaduran plains. But the enthusiasm and the mass departure of men, the exploration and settlement of new lands, the mingling of races, and the rapid transfer of culture would have been entirely different had it not been for the pursuit of Indian treasure and gold and silver mines.

Three centuries had to pass after the Spaniards and Portuguese had explored all the lands of Middle America, New Spain, and South America before the Anglo-Saxons crossed the Mississippi and reached the coast of California. The incentive for their success was the discovery of gold nuggets in the sands of rivers in California. Apparently, the desire for gold also motivated English-speaking men and led them to settle almost half of the present United States. If the gold deposits of California had been discovered earlier, California, Oregon, Nevada, New Mexico, and Arizona might still be Mexican states, for Mexicans would have flocked there as eagerly as the Anglo-Saxons did later.

Certainly in the course of occupation of Ibero-America there were episodes involving gold and silver which deserve censure. Among the most notorious and controversial events are the torture of Cuauhtémoc, the last Indian emperor of Mexico, in an attempt to discover his treasure, and the seizure of Atahualpa's treasure in

Cajamarca on the pretext that it was his ransom. But these were isolated events that took place during the heat of the Conquest. However lamentable they may be, they cannot be considered as characteristic of the transcendental accomplishment of discovering, occupying, and settling a vast continent and creating many nations. Salvador de Madariaga has commented that Spain committed as many errors and was as guilty in its actions in the Americas as could be expected from any nation of humans in that period and in view of the ideas then current concerning conquests and the rights of victors. In addition, he points out, we must also recognize the circumstances of distance and the conditions of adversity in which the conquerors found themselves.

What is new and surprising is the fact that the conquistadors' contemporaries and the Spanish Crown itself were deeply aware of these excesses. Furthermore, there are the examples of outstanding missionary work such as the labor of the twelve Franciscan friars who came to New Spain at Cortes's request and the good work of Vasco de Quiroga, a magistrate of the *Audiencia,* or High Court, of Mexico. Quiroga, ordained a priest, was consecrated a Bishop and spent many years of his life in the province of Michoacán where he founded a "hospital" which cared for the ill, the infirm, and the aged and in which he and his followers taught arts, crafts, and trades to raise the level of existence of the natives. To this day his name and memory are venerated by the Tarascan Indians and by all Mexicans.

The most prominent and best known, but not the only accusing Spanish voice of "holy anger" (the "roaring noise," as Agustín Yañez expresses it),[1] among those who called into question and debated the Crown's right to take over the Indies and the Spaniards' actions in the newly discovered countries was Fray Bartolomé de las Casas, Bishop of Chiapas, appointed Protector of the Indians by Charles V. For more than fifty years Las Casas battled for his cause: the freedom and rights of the natives of the New World. Las Casas's influence with the King and the Council of the Indies was so great that he was permitted to make thirty trips between Spain and America, had freedom to publish his works of devastating criticism (among them one used most effectively by Spain's European enemies, entitled *The Destruction of the Indies*), and decisively influenced the promulgation of laws protecting the Indians.

Because of its obvious parallelism, Ortega y Gasset's comment in regard to Rome's conquest of Spain and the razing of Numantia by Scipio Aemilianus would not be out of place here. He said: "Let us not take offense at the outrage they inflicted on our ancestors. Without it, and others like it, the soul of Spain would not be what it is, and would not have that solid foundation of Roman stonework which is always present in the Spaniard."[2]

Many Anglo-Saxon historians and writers, however, still persist in viewing the Conquest and colonization of the New World as having taken place under two vastly different sets of circumstances. The Hispanic peoples, they contend, came only for plunder and rapine, quickly returning home with their ill-gotten gains. In contrast, they see the English colonizers as having come with their families to build homes, till the soil, and leave farms to be inherited by children who would inter their parents' bones in quiet New England churchyards. But the reality was not quite so. Englishmen murdered Indians for their lands and meager possessions. They justified their actions with the excuse that the Indians were bloodthirsty savages who made poor use of their land and killed capriciously and therefore deserved to be killed first so that diligent Europeans could make full use of the neglected soil.

The North American historian Lewis Hanke appreciates these facts, for he writes in his book *The Spanish Struggle for Justice in the Conquest of America* that:

> the Spanish conquest of America was far more than a remarkable military and political exploit; . . . it was also one of the greatest attempts the world has seen to make the justice of Christian precepts prevail in one brutal and sanguinary time.

The thesis of this book is that the clash of arms was not the only struggle during the Conquest. The clash of ideas that accompanied the discovery of America and the establishment there of Spanish and Portuguese rule is a story that must be told as an integral part of the Conquest, and endows it with a unique character worthy of note. Of course, many nations have had a "habit of acting under an odd mixture of selfish and altruistic motives," as Woodrow Wilson expressed it when describing the history of the United States. No European nation, however, with the possible exception of Portugal, took her Christian duty toward native peoples so seriously as did Spain. Certainly England did not, for as one New

England preacher said, "The Puritan hoped to meet the Pequod Indians in heaven, but wished to keep apart from them on earth, nay, to exterminate them from the land."

The Spaniards, or at least many of them, had an entirely different attitude toward the Indians and the desirability of incorporating them into a Christian and European civilization, as the disputes on their character have shown. Those loud and dogmatic voices quarreling over the Indians have a peculiarly human ring, representing as they do opposed theories of human values as well as philosophical and theological differences.[3]

There is no denying that the style and character of the Hispanic and English occupation of the New World were quite different. But it is doubtful that the differences can be ascribed merely to a "flaw" in the character of one nation or another, for that would have us assume that *all* the people of any nation are perfect spiritual reproductions of one another. Historians of English colonization can point to good men such as William Penn and Roger Williams and to others such as Lord Jeffrey Amherst—who gave blankets infected with smallpox to Indians as presents—while Spain can offer evil men such as Nuño de Guzmán and Alonso Aguirre and good men such as Padre Las Casas and Bishop Vasco de Quiroga. The Portuguese can contrast the work of Mem de Sá and Padre Antônio Vieira with that of Raposo Tavares and Domingos Jorge Velho. Differences can more rationally be ascribed to the rapid changes in European life in early modern times and the nature and culture of the land and people who were conquered. Frederick A. Kirkpatrick in his book *The Spanish Conquistadors* recognizes these influences when he states:

> In viewing the work of Spaniards in America, thought naturally turns to the later work of the English farther north. Points of contrast at once occur. Since the first permanent Spanish settlement dates from 1493 and the first permanent English settlement from 1607, both countries reproducing themselves in the New World, the England so reproduced was the England of the Stuarts and the Commonwealth, whereas the Spain so reproduced was that of the Catholic sovereigns and of Charles V. Spanish settlement coincided with the period of adventurous exploration: English settlement followed the period of adventure. When the Spanish Conquistadores are accused of inhumanity and inefficiency, this difference of time must be remembered: all that has been said—in the first

> instance by Spaniards—about that inhumanity and inefficiency is true, but not the whole truth. It may be noted that during the same period the English too were pursuing conquest and colonisation in Ireland: and one would hesitate to claim that their work was more efficient or more humane.
>
> Thus the two movements differed in the world which they brought with them; they differed still more in the world which they found: the English found no Mexico, no Peru, no Bogotá.[4]

Today it cannot be denied that European expansion must be reexamined in the light of new knowledge and new attitudes: each generation must rewrite history. We can, therefore, appreciate that European expansion in the sixteenth or nineteenth or twentieth century always smacked of the horrors associated with the term *imperialism.* Natives have always suffered and died; some Europeans have become wealthy. But in the process changes do take place, and we would have to be intellectually and morally blind not to recognize that most of the changes are inevitably in the direction of promoting a new culture. When people exchange primitive ways for modern technology and medicine and blend their indigenous cultural and literary heritage with one another, a richer and better life must eventually emerge.

Notes

1. Agustín Yañez in a course of lectures on Padre Bartolomé de Las Casas offered in the Colegio de México in September–October 1966 for the fourth centenary of the famous Dominican's death.

2. Ortega y Gasset, *Conde de Yebes: Veinte años de caza mayor* (Madrid: Espasa Calpe, S.A., 1943), p. xxii. Also in his *Obras completas,* 3d ed. (Madrid: Revista de Oriente, 1955), p. 431.

3. Lewis Hanke, *The Spanish Struggle for Justice in the Conquest of America,* pp. 1 and 175.

4. Frederick A. Kirkpatrick, *The Spanish Conquistadores* (London: 1934), ch. XXVII.

III

Mining: Motive Force of the Discovery

Directing our attention to the work of the miners in America, we can say that ill-gotten treasure did not interest these men. They wished to locate and work the very mines themselves. This drive carried them over some of the most difficult obstacles of nature, resulting in the exploration and opening to Western civilization of the farthest reaches of the Indies. Given the incredibly difficult geography of the hemisphere, only the desire to discover the sources of precious metals could induce these men to scale precipitous mountains and cross inhospitable plains. Only against the perspective of the geography of this region can the work and success of the conquistador and miner be fully appreciated.

The American continents have as their axis the ranges of mountains which run along the west coast of South and North America and through the center of the isthmus of Middle America. Starting in the south at the island of Tierra del Fuego, the high and rugged Andes run northward close to the Pacific shore of South America virtually to the border between Colombia and Panama. This mountain chain—or *cordillera,* as it is called in Spanish—is generally long and narrow, but it broadens out at the Altiplano of Bolivia to some 400 miles. Just north of the Ecuadorian-Colombian border it splits into three ranges: the western range runs along the Pacific coast to end at the border with Panama; the central range runs due north to the Caribbean at Baranquilla and Santa Marta; the eastern range runs south of Lake Maracaibo and then cuts north to the Caribbean coast of South America where it continues eastward, finally descending to sea level just west of the delta of the Orinoco and the island of Trinidad.

A sixteenth-century engraving from Agricola's treatise on mining could almost serve to illustrate gold-mining practices in eighteenth-century Brazil. The small hoelike implement (E) is similar to the Portuguese miner's pick, or *almocafre,* and the catch basin (H) is clearly related to the *carumbé,* while the oval bowl for washing pebbles (I) resembles the Brazilian *bateia.* From Georgius Agricola, *De re metallica* (1556). (*Rare Book Division, The New York Public Library, Astor, Lenox and Tilden Foundations.*)

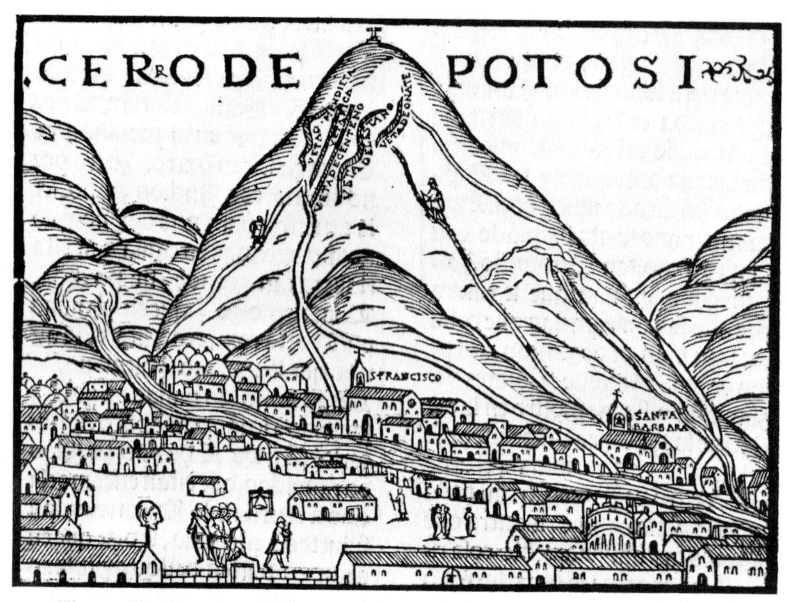

The earliest published view of the city and great silver mountain at Potosí, Bolivia, appeared in 1553 in *Crónica del Perú* by Pedro de Cieza de León. (*The Hispanic Society of America.*)

A second range, the Central American, runs from central Panama, just north of the Canal, through the middle of Central America to the Isthmus of Tehuantepec in Mexico, where it falls away leaving a transisthmian gap. North of Tehuantepec is the complex end of the Rocky Mountain chain which stretches north to Alaska. The Mexican end of this chain is marked by the low-lying Chiapas Highlands, the jumbled topography of the Sierra Madre del Sur, the rugged Sierra de los Volcanes near Mexico City from which the great Central Plateau of Mexico extends northward to the Great Basin of the United States. The so-called Eastern and Western Sierra Madre are really the dissected edges of the Central Plateau. To the west of the Mexican mainland, across the Gulf of California, lies the peninsula of Lower California, a dry, mountainous, mineralized finger of land jutting into the Pacific.

Three other mountainous areas should also be mentioned. First, there are the Brazilian Highlands which occupy much of east-central Brazil. These mineralized, low-lying mountains have, paradoxically, acted both as a barrier to penetration of the Brazilian interior and as a magnet to explorers as their wealth of gold and diamonds was disclosed by the *bandeirantes,* Brazil's pioneer trailblazers of the backlands. Second are the Guiana Highlands in the northeastern corner of South America west of the delta of the Orinoco. These relatively low mountains, covered by dense tropical vegetation, have been immortalized in William H. Hudson's famous romance, *Green Mansions.* Although the mountains are difficult to penetrate, discovery of gold in the region brought in numerous prospectors and almost caused a boundary war between Venezuela and British Guiana. Further discoveries of iron ore, bauxite, and industrial diamonds keep attracting explorers despite the great obstacles. The third mountainous region is that of the Greater Antilles, geologically related to the Central American range. Although its gold has largely been depleted, the region yielded enough treasure at the beginning of the Spanish penetration to entice the Catholic Monarchs' subjects to further explorations. Today these islands are still storehouses of industrial minerals, particularly iron ore, bauxite, and nickel.

But it is the great north-south axis of mountain ranges that constitutes the geological, historical, and social raison d'être of Spanish America, although at the same time it was also the chief obstacle to the region's exploration. Despite the hardships presented by the mountains, the settlements in this region became centers for life and culture, leading eventually to the formation of many distinctive peoples and nations. Only the eager search and fortunate discovery of precious metals could work the glorious miracle of bringing the Spanish-speaking American countries into existence.

One needs to travel over the region himself, even if it be by plane, to appreciate the accomplishment of the early explorers. Travel over the Andean region from Colombia to Chile; cross the mountainous tangle of Bolivia; traverse the mountain ranges from west to east, from Lima to Rio de Janeiro, and from Santiago to Buenos Aires, and you will grasp the significance of the prodigious, almost incredible human effort that was required to cross those lands on foot or on horseback. Some of the areas are rocky, many are sandy wastes, almost all are steep and without vegetation, and in many places rainfall is unknown. Look at this rugged land and then reflect on the fact that these heights were all explored by Spaniards and Portuguese, who founded cities in them, many at altitudes of over 9,000 feet, and constructed roads to connect these urban centers.

The hardships overcome by the intrepid Spaniards were graphically portrayed by Pedro Cieza de León, who spent seventeen years in South America tramping over the terrain traversed by the conquistadors:

> In this land of Peru there are three desert ranges where man can in nowise exist. One of these comprises the forests of the Andes, full of dense wildernesses, where men cannot nor ever have lived. The second is the mountainous region, extending the whole length of the Cordillera of the Andes, which is intensely cold, and its summits are covered with eternal snow, so that, in no way, can people live in this region, owing to the snow and the cold, and also because there are no provisions, all things being destroyed by the snow and by the wind, which never ceases to blow. The third range comprises the sandy deserts from Túmbez to the other side of Tarapacá, in which there is nothing to be seen but sandhills, and the fierce sun which dries them up, without water, nor heat, nor tree, nor created thing, except birds, which, by the gift of their wings, wander wherever they list.[1]

In later years, Don Fernando Belaunde, the ex-President of Peru, who also knew the terrain of his nation, called the work of the conquistadors "cyclopean" and noted that they faced continual mysteries, hardships, struggle, and even death. Yet they prevailed and built their baroque cities and churches, their aqueducts, bridges, and roads. The search for mines was the motivating force, and today there is scarcely a mine in the Andes without traces of the work done by the Spaniards.[2] Only mining, the search for metals hidden in the bowels of these forbidding mountain ranges, can explain such achievements.

Mountainous terrain was not the only obstacle to rapid exploration of those recently discovered territories. Other and still greater barriers existed, many of which have not yet been surmounted: the jungles, the marshes, the swampy basins of the great rivers on the eastern slope of the Andes. These regions would have remained unknown and unexplored had not the Spaniards been attracted by gold and silver.

With but few exceptions, the lowlands and plains of Ibero-America also are hostile to man. We think of the pampas, the fertile Central Valley of Chile, the lovely mountain basins of the Andes and Mexico, and the coffee lands of São Paulo and Paraná as "typical." Yet they are quite atypical. Far more characteristic are the tropical savannahs of the Chaco, the Llanos of Venezuela and Colombia, the flood plains of the Amazon, the lagoon coast of the Caribbean, the barren strip of land on the western coast of South America between the cordillera and the sea, and the desert of northern Mexico. Even the pampas of Argentina and Uruguay defied human habitation and exploitation until the invention of the railroad, the steel plow, and barbed wire. Only the strongest of motives—the search for mines—could induce men to open these areas and erect a civilization based, paradoxically, on wealth brought to the surface from mineral deposits in these hostile lands.

"Neither before nor after, neither in Spain nor in the rest of Europe, were there bolder or more singular persons than the Spanish exporers and conquistadores," the Colombian scholar Germán Arciniegas has said.[3] They discovered and explored vast territories under the stimulus of a legend which told of a land abundant in gold beyond the dreams of avarice, a land where, as a sacrifice, the Indian chiefs threw gems and objects of gold into a lake in which their king dove entirely covered with gold dust. This was El Dorado, the Golden Man. Despite its ingenuousness, says Padre Bayle, "there was no lure in all America, nor in all the world, that aroused so many desires, or was the goal of more expeditions, or the cause to which were sacrificed more pains, money, and men" than this legend of El Dorado.[4] With so many failures and so many victims, the legend spurred the discovery and exploration of the whole basin of the 1,700-mile-long Orinoco and its tributaries, from the jungles of Venezuela as far north as the Guianas and west to the river's Colombian tributaries, up to the region of Cundinamarca. And these intrepid men explored as well the entire course of the Amazon, more than 3,900 miles long, and its network of equatorial tributares.

One of the expeditions which resulted in the exploration of the course of the Orinoco was under the command of Diego de Ordaz, one of Cortés's captains. He is the man who, it is said, was the first to climb Popocatépetl to gather sulfur in its crater in order to manufacture gunpowder. The house he occupied in Coyoacán, a town in the state of Mexico, is still in existence, near Cortés's own house. It does not appear, however, that he enjoyed it for long, nor did he rest on his laurels, for we find that in 1531 he sailed from Sanlúcar de Barrameda in Spain with three well-provisioned ships and precise instructions from the King. In pursuance of the royal orders, he carried out the exploration of the Orinoco from its mouth to well inside its tributary, the Meta, which then began to acquire its great renown within the complicated lore of El Dorado. In memory of this great explorer, Puerto Ordaz, a town recently founded by the Compañía Siderúrgica del Orinoco (Orinoco Steel Company), has been named in his honor.

The exploits of Francisco de Orellana, a native of Trujillo, Spain, were also performed under the spell of the elusive El Dorado. They can be described as among the most extraordinary imaginable. Arciniegas says of him: ". . . that man who, by traveling from Guayaquil—on the Pacific—to the mouth of the Amazon [on the Atlantic], made a dramatic journey, perhaps infinitely longer and more hazardous than Columbus'."[5]

A search for riches was also responsible for the Portuguese exploration of the continent. Aleijo Garcia, a survivor of the Juan Díaz de Solís expedition to the Río de la Plata in 1516, led hundreds of Guaraní Indians across Brazil and Paraguay in 1524 toward a fabled "mountain of pure silver" and is said to have passed within a hundred miles of the then undiscovered Potosí, only to perish when the party was defeated by other tribes.[6] In general, however,

Portugal did not have as many heroic figures in the New World as Spain, nor could such figures really emerge without native kingdoms to conquer. The courageous and resourceful bandeirante, prospecting for precious stones and gold, was the nearest equivalent to the conquistador.

One of the most remarkable bandeiras was that of Antônio Raposo Tavares, which lasted from 1648 to 1651. This intrepid explorer led a group of 200 Portuguese and mamelukes, accompanied by more than 1,000 Indians, through the interior on one of the longest treks ever undertaken in the hemisphere. The bandeira left São Paulo and headed down the Tietê River to a spot where Corumbá now stands, moving then through the São José range to the marshes of Izozog and the region of the Serrano Indians. From there it reached the Paraguay River and proceeded by land to the Jesuit *reducciones* of nearby Mboimboi and Maracajú. When they met armed resistance, the bandeirantes paddled up the Paraguay to the Guaporé. Jaime Cortesão, the Portuguese historian and geographer, believed that Tavares led part of the group overland to the Andes, reaching the city of La Plata and then traveling back to the Amazon basin and so on down the Madeira to Belem do Pará. If we include the Andean incursion, the whole trip covered about 7,500 miles; if we consider the shorter route only, which is definitely documented, the trip covered some 6,250 miles. It is understandable why this is sometimes called "the greatest bandeira of the greatest bandeirante."[7]

Notes

1. *La Crónica del Perú* (1553), chap. XXXVI, in C. Markham, trans., *The Travels of Pedro Cieza de León,* Hakluyt Society, ser. I, no. 33 (Cambridge, 1864), p. 128.

2. Interview granted to the Spanish journalist Luis Calvo and published in *ABC,* Madrid, on June 20, 1963.

3. Introduction to Miguel Albornoz, *Orellana: El caballero de las Amazonas* (Mexico: Editorial Herrera, S.A., 1965).

4. Constantino Bayle, S. J., *El Dorado fantasma* (Madrid, 1943), p. 18.

5. Introduction to Albornoz, *Orellana.*

6. Lewis Hanke, "The Portuguese in Spanish America, with Special Reference to the Villa Imperial de Potosí," *Revista de Historia de América,* vol. LXI (June 1961), p. 5, and Charles E. Nowell, "Aleijo

Garcia and the White King," *Hispanic American Historical Review,* vol. XXVI (1946), pp. 440–446.

7. Jaime Cortesão, *Rapôso Tavares e a formação territorial do Brasil* (Rio de Janeiro, 1958). See also his article "A maior bandeira do maior bandeirante," *Revista de História,* vol. XXII, no. 45 (January–March 1961), pp. 3–27 and excerpts, translated in Morse, *The Bandeirantes* (New York, 1965), pp. 100–113.

IV

Mining: The Instrument That Forged America

Let us now turn to work of the miners, those rare and complex men who blended intuition and skill, technical competence and perseverance, imagination and caution, who were determined, courageous, patient in adversity, vital and generous in success, and always deeply in love with their profession. And then let us say something of their progress through American lands.

When we speak of *miners,* we generally understand the word to refer to the mine operators and promoters, the technicians (engineers and the like), and the men who labor to extract minerals from the earth. But the term also ought to include the prospector, called a *gambusino* in Mexico and a *faiscador* in Brazil. Throughout the colonial period the prospector was one of the chief protagonists in mining. The gambusino possesses the virtues characteristic of miners, including the highest degree of enthusiasm and zeal for travel and the discovery of mines. What distinguishes a gambusino is that his grasp of the subject—for it is not easy to recognize and pick out metalliferous rocks from among the others—has not been acquired in lecture rooms or in books but in life itself, in the midst of nature, by traveling through the hills, through scrub and desert, seeking out and observing strata, inclines, stones, patterns, colors, and even odors which may indicate to him the location of metals and useful substances. With very few exceptions, the heroic period of Spanish American mining—the century following the discovery of the New World—was in the hands of men who, endowed with the physical resistance, enthusiasm, and pertinacity for such excursions, possessed this almost-divinatory gift for finding metals not only in their natural state in placer deposits but also in rocks.

The first allusion to Spanish miners in America is made by Fray Bartolomé de Las Casas when he says that Columbus arrived in Hispaniola in 1493, on his second voyage, with tools and with 1,500 men, among them "a large proportion of laboring folk . . . to take gold out of the mines."[1] The following year, the Monarchs were petitioned to send gold washers and miners from Almadén "to gather it in the sand, and the others to dig for it in the ground." This undertaking apparently was not successful, for Las Casas tells us that "none of them was found to prosper." Gold mines were opened in Cuba with more success, and placer deposits were exploited in Puerto Rico, but Cortés said of them in 1526 that "before such trade was carried on [with New Spain] there was not a man in the Islands [the Antilles] who owned more than a thousand pesos of gold."

As for the mainland, we know that when Columbus reached the shores of Veragua (the western part of Panama) on this fourth voyage, he found indications of gold deposits which had been very briefly worked, and which never attained importance; this was also the case of the investigations made on the Gulf of Darién and in other parts of Central America.

True mining began in the lands discovered along the Gulf of Mexico, in what was soon to be called New Spain, and properly so, for Old Spain was destined to pour her best into that region. Evidence of the existence of gold and silver deposits was collected by the expeditions of Francisco Hernández de Córdoba in 1517 and Juan de Grijalva in 1518, both of which reconnoitered the coast of Middle America before the Conquest, and by the expedition of Cortés himself in 1519. Andrés de Tapia, a member of Cortés's expedition, tells us that on the coast of San Juan de Ulúa, Veracruz, Cortés received a gift from Montezuma II consisting of "one gold wheel and another of silver." Each was as big as a cartwheel, though not very thick, he reports, and "they say were made to represent the sun and the moon."[2] Cortés lost no time after the Conquest in sending explorers to different and distant territories where he had been informed of the existence of precious metals and where, as Bernal Díaz tells us, "According to the books of Montezuma [Montezuma II] we saw from whence they brought the tributes of gold, and where there were mines." It is certain that the first Spaniards to work gold mines in New Spain on a permanent basis were those whom Cortés sent "to the matter of Tustepeque." Bernal Díaz even gave the names of the miners who stayed behind to work.[3]

The first silver mines exploited by the Spaniards in New Spain were those of Morcillo in the present state of Jalisco in about 1525, but they were soon abandoned as were those of Villa de Espíritu Santo near Compostela, Nayarit, at about the same period. A few years later in 1530 the Spaniards opened the workings of Zacualtipán and Sultepec in the present state of Mexico and at Zumpango in Guerrero. The silver lodes of Tasco, Guerrero, were first worked in about 1534, and a few years later those of Tlalpujahua, Michoacán, and Amatepec, Mexico, also were worked.[4] As Cortés himself informs us, tin and copper were extracted in Tasco before the opening of silver mines there.[5] With these metals he manufactured bronze for cannon. The first silver miners in Tasco were members of Cortés's expedition. Cortés himself exploited mines in Tasco in the Cantarranas district, "with its houses and churches and three workshops: an ordinary water mill for crushing ore with eight mallets, another with vats and grinders for washing ore. Another for retreating ores, with its wheel and other necessary gear."[6]

The first adit (tunnel or gallery) opened by the Spaniards in the continental New World was the so-called Cortés tunnel in New Spain, in Tehuilotepec in the neighborhood of Tasco. This gallery was some 90 meters (300 feet) in length and so large that it could be entered by a man on horseback. The use of a pump to draw the water out of these mines in Tasco is also attributed to Cortés.

The chief mining camps, or *minerales,* as the mining centers are called in Mexico, opened in rapid succession. Among the mines that became world famous or are still being exploited, many were opened during the first thirty-five or forty years after the Conquest. In 1546, Juan de Tolosa discovered the well-known mines of Zacatecas at the foot of the Cerro de la Bufa near the site of the present city of Zacatecas. After a few exploratory operations, exploitation of the first mine began in 1548. Although their long history includes many natural fluctuations in the level of activity, the mines are still in operation. The most important lodes of the Guanajuato mines were discovered between 1548 and 1558; in the latter year the famous "mother lode" became known. Together with "La Valenciana," this deposit was to spread the name of that much-admired city throughout the world. In 1547, the "Santa Bárbara" was discovered, the first in the present state of Chihuahua, at a distance of more than 1,200 miles from Mexico City. Five years later, no more than 60 miles from Mexico City, the mines of Pachuca and Real del Monte were discovered. It was there at the "Purísima Grande" mill that Bartolomé de Medina of Seville invented and tested his famous method of amalgamating silver ores in 1554.

In 1552 an event of great importance for the present-day Mexican iron and steel industry occurred: the discovery of the iron mines of Cerro de Mercado. (A short time later Francisco de Ibarra founded to the south the city of Durango, capital of Nueva Viscaya.) The discovery of this very important iron deposit was due to a curious error. The Royal Audiencia of the Kingdom of New Galicia, a tribunal charged with the administration of north-central and northwest Mexico, received information about a mountain of silver containing enormous riches and sent Ginés Vázquez de Mercado, a nephew of Ginés Vázquez de Tapia, one of Cortés's captains, to investigate the matter. His disillusionment was great when he saw that what had been thought to be an immense mass of silver was a mountain of iron ore. Returning from his mission, the discoverer of the mountain, which still bears his name, was seriously wounded by some Tepehuan Indians near Sombrerete, in the modern state of Zacatecas, and he died of his wounds soon after. The Cerro de Mercado, the rights to whose ore were acquired by the Compañía Fundidora de Fierro y Acero de Monterrey, S.A. (Monterrey Iron and Steel Co., Inc.), in 1920 from an American estate, has been to the present day the chief source of ore for the blast furnaces producing pig iron for the Mexican steel industry.

In 1553 Francisco de Ibarra discovered the mines of Fresnillo, in the present state of Zacatecas, which are still being fully exploited today. At about the same time, Francisco de Urdiñola, one of Cortés's sergeants who later became the Marquis of Aguayo, discovered the mines of Mazapil, located on the border between the present states of Zacatecas and Coahuila. The mines of Sombrerete and Chalchihuites, also in Zacatecas, as well as those of Temascaltepec, in the state of México, have been known since 1555.

A number of important towns in New Spain owe their origin to mining, and most of them are linked to the names of colorful and important men. The aforementioned Tasco in the state of Guerrero had its most splendid period between 1748 and 1757 with the bonanza of the mines of Don José de la Borda. During this time Borda built the beautiful church of Santa Prisca. It was long believed that Don José was French, but the Mexican historian Manuel Toussaint has offered convincing proof that the famous miner was a Spaniard born in 1699 in Jaca in the kingdom of Aragón.[7] He was a skillful, persistent, and generous miner who confronted many business reverses with great integrity. Originally Borda made a fortune in a great bonanza in Tlalpujahua in partnership with Don Manuel Aldaco. Aldaco, a native of Guipúzcoa, together with Don Francisco Echeveste, another Guipúzcoan, and Don Ambrosio de Meave, of Durango, Biscay, founded in Mexico City the Colegio de las Vizcaínas (the Basque School) which is still in existence and is a pride of the Mexican capital because of its excellent organization and beautiful architecture.

After the Tlalpujahua bonanza, Borda went bankrupt and begged a loan from the Archbishop of Mexico. The Archbishop, remembering Borda's magnificent beneficences to the church, gave him a gold candelabrum—which Borda in more affluent days had presented to the cathedral at Mexico City. Just before the money was expended, Borda's workers struck the Tasco bonanza. Eventually the Tasco mines were exhausted and Borda was again ruined, only to recoup his fortune operating mines in Zacatecas.[8]

Borda was held in high esteem by his contemporaries. The eminent jurist Don Francisco Xavier de Gamboa wrote of him in 1761:

> And Don José de la Borda has always cast luster, for he may be considered as the foremost miner in the world owing to his vast comprehension and great accomplishments in this field, both in association with his brother Don Francisco and alone, working in various mines, alternately experiencing prosperous and adverse fortune; and most recently in the mines of Tasco, where he has built and so liberally and magnificently endowed the parish church, repository of very rich jewels, ornaments, sacred vessels, and all the service of the church in gold and silver, [and] whom Pope Benedict XIV by his Brief issued in Rome on the fourth day of March, 1754, covered with praise and blessings for such commendable works offered to God, the veneration and ornamentation of His temples.[9]

Gamboa went on to relate the opinion of Manuel Antonio Rojo, Archbishop of Manila, who:

> has written to the King after having admired so rare and imposing a work when he went to offer the first rites in that temple, before leaving for his Archbishopric [naturally, we might add, on his way to Acapulco to take ship], expressing himself as follows in his letter of the fifteenth of March, 1750: "I paused in Tasco for the ceremonies of the dedication of the parish church [Santa Prisca], a magnificent work by Don José de la Borda: in the perfect and beautiful architecture; in the completeness of its decorations and the richness of its precious furnishings, I doubt there be another like it in all of Christendom. What he has spent amounts to a million; but his heroic piety and humility are even greater than his vast works: there is not a single tablet nor jewel which shows the slightest indication that he is the benefactor; but they publish the fact by their magnificence."[10]

Guanajuato also attained its greatest renown during the second half of the eighteenth century with the exploitation and wealth of the La Valenciana and Rayas mines which so greatly impressed Alexander von Humboldt. La Valenciana was worked by Antonio Obregón y Alcocer, a Mexican-born grandson of a native of Santander. He eventually became Count of La Valenciana and was later associated with Pedro Luciano Otero and Diego Rul. The operations carried out in these mines, which produced an estimated 300 million dollars in fifty years, have been the object of praise by many writers including Humboldt, who visited them and dedicated a large section of his long chapter on mining to them in his *Essai Politique.*[11] Over the mother lode (*veta grande*), Obregón opened the shaft known as El Santo Cristo de Burgos, 493 feet in depth, and later the hexagonal Nuestra Señora de Guadalupe, which reached a depth of 1,130 feet. Lastly he dug the octagonal general shaft called Señor San José, with a perimeter of 88 feet and an eventual depth of 1,685 feet. Humboldt called this one of the greatest and boldest undertakings in the history of mining. "In 1767 he [Obregón] went into partnership with a small businessman of Rayas named Otero," says Humboldt. "How could anyone have expected then that in the course of a few years, he and his friend would be the richest individuals in Mexico and perhaps in the whole world?" And he adds that Obregón "preserved, in the midst of immense wealth, that simplicity of habits and that openness of character which distinguished him in less fortunate times."

Mining activities in the city of Nuestra Señora de las Zacatecas continued at a peak level for nearly two centuries. In 1629 it was described as the third city of New Spain after Mexico City and Puebla de los Angeles in population, business activity, and wealth.[12] Its fame was celebrated in 1718 by Presbyter Don Juan de Santa María Moraver:

> This most rich city is so well-known by that epithet, that as many mouths publish it abroad as open mines proclaim it with tongues of silver, the yearned-for fruits of its rich metals softening the harshness of its mountains, so great being the quantity of silver given us by Divine Omnipotence that all tread upon it without reflection; the whole city is situated upon veins of it, for scarcely does Heaven bedew it with rain, than all is uncovered in lodes and veins that announce it; this is proclaimed by its ample shipments, its constant works of charity, its vast expenditures, and its repeated generosity. . . .[13]

The people of Zacatecas showed their gratitude for the heavenly beneficence by erecting one of the most beautiful Churrigueresque cathedrals in New Spain.

The mines of Pachuca and Real del Monte in the state of Hidalgo are important producers of silver to the present day. The greatest mines were operated by Don Pedro Romero de Terreros, who became a member of the Order of Calatrava and Count of Regla. He distinguished himself by his many charitable actions and gifts to the Church and to missions. Humboldt says that the Count of Regla "had built in Havana at his own expense, in mahogany and cedar, two ships of the line of the largest size, which he presented to his sovereign," and that his youngest son, the Marquis of San Cristóbal, "distinguished himself in Paris by his knowledge of physics and physiology."[14]

Monterrey, the present capital of the state of Nuevo León, was founded in 1579 by Don Luis de Carbajal de la Cuva, a *marrano,* or converted Jew, but it soon declined and did not rise in importance again until it was reestablished by Diego de Montemayor. The nearby mines were consolidated by General Agustín de Zavala, a miner from Zacatecas with funds and workmen from that city.[15] In 1900, the first blast furnace using metallurgical coke, as well as the first combined steel works in Spanish America (Compañía Fundidora de Fierro y Acero de Monterrey, S.A.), were established in Monterrey.[16]

Although the city of Sombrerete in the state of Zacatecas is in decline today, it has had periods of spectacular prosperity. Humboldt wrote of the most important mine of the district:

> The Fagoaga family, known for its charitable works, its enlightenment, and its zeal for the public welfare, presents the example of the greatest wealth ever furnished its owners by a mine. A single vein owned by the family of the Marquis of Fagoaga in the district of Sombrerete, produced in five or six months a net profit of 20,000,000 francs.[17]

Similar stories could be told of the cities of San Luis Potosí, Oaxaca, and Guadalajara which also owe their founding or prosperity to mining.

Obviously the men who accompanied Cortés to Mexico and their immediate successors did not rest on their laurels, for in a very short span of years they traveled over all the territory of New Spain. Throughout the Viceroyalty and close by the mines they established the first important towns as an outgrowth and development of the *reales de minas,* the centers which grew up around the mining establishments. Indeed, the term *real de minas* now has become restricted to workings of gold and silver.

The southward colonization of the continent followed the same pattern—one new expedition after another was organized, stimulated by real or legendary reports of wealth in metals such as El Dorado. Expeditions were also made from the northern coasts of Colombia (Castilla de Oro) and Panama (Darién) toward the interior of Colombia and to Peruvian shores, and from Río de la Plata toward the northwest of Argentina.[18]

In August 1535, the ill-fated expedition of Don Pedro de Mendoza left Spain to establish a settlement, Nuestra Señora Santa María de Buenos Aires, on the banks of the Río de la Plata (the River of Silver) at the southern end of the continent. The settlement proved a disaster: there were no mines, few Indians to place in *encomiendas,* and little food. Starvation dogged the colonists, and when Mendoza returned to Spain in 1537 for help, he left Juan de Ayolas as his Lieutenant-Governor. Ayolas, meanwhile, had sailed up the Paraná River where he found friendly Indians and founded the city of Asunción. Leaving a small garrison behind and notifying Buenos Aires of his doings, he plunged deeper into the continent, striking out for the source of the silver which had filtered eastward from the Andes to give La Plata its name and reputation. Ayolas and his men reached their destination and started back with a cargo of silver, only to be ambushed. Ayolas's deputy, Martínez Irala, then moved the original settlement from Buenos Aires to Asunción, where the colonists recuperated and even prospered in their isolated bucolic retreat.

Among the outstanding episodes in the history of the discovery of the Río de la Plata was the exploit of Alvar Núñez Cabeza de Vaca, the first European to behold that marvel of nature, the Iguazú Falls, at the point where Brazil, Argentina, and Paraguay come together. This man was the same extraordinary traveler who, after misfortunes and shipwrecks, successfully undertook an eight-year journey on foot for more than 6,000 miles from Florida across Texas and Northern Mexico to Mazatlán on the Pacific coast of New Spain. After the King had rewarded his famous exploits by appointing him Captain-General and Governor of Río de la Plata, he was willing to leave the three ships of his expedition and undertake a cross-country journey of more than 600 miles through dense jungle. Traveling on horseback, on foot, and in Indian canoes, his expedition journeyed from the Brazilian coast near the Ilha de Santa Catarina to the confluence of the Iguazú and Paraná Rivers. And with that complete naturalness with which these men accomplished their epic journeys, he says in his account that he was obliged "in order to avoid the obstacle of a cataract in the river to carry the canoes overland for a league." That obstacle was no less than the spectacular Iguazú Falls! Alvar Núñez went up the Iguazú and Piquerí Rivers, where he believed silver mines were to be found. As the scribe Pero Hernández says, "Because of the nature of the land, it is held to be certain that if silver mines exist, they are to be found there."[19] But silver was not to be found. Although the mines proved to be illusory, they were the spur to filling in another large space on the map of South America.

On the Pacific side of the Andes, the explorations of Pascual de Andagoya in 1522 and Pizarro's expeditions in 1524 and 1532 were significant. On the Isla de la Plata, Pizarro and his thirteen companions found some silver and a few pieces of gold jewelry, from which they inferred that Andagoya's report of riches in *Birú,* or *Pirú,* was correct. The gifts Pizarro received from the natives of Túmbez on the Gulf of Guayaquil made it possible for him to present gold and silver vessels to the King and Queen of Spain in 1528 when they agreed to his terms for the conquest of Peru. Pizarro's second expedition started for Cajamarca and Cuzco in

1532. The Spaniards were astounded by the riches of the temples and houses of the royal family where gold and silver had been collected. With the "Ransom of Atahualpa" and the spoils of the famous treasures of the Peruvian cities and their surrounding districts, the Spaniards saw their ambitions for gold and silver more than fulfilled: more than being a mere indication, the presence of these treasures was undeniable proof of the existence of fabulous deposits of precious metals.

Another land of treasure was opened by the expeditions of the brothers Gonzalo and Hernán Jiménez de Quesada in 1536 and Juan Vadillo in 1537 when they traveled south from Cartagena on the Caribbean coast of modern Colombia, along the Magdalena and Cauca Rivers. At the same time, the expeditions of Sebastián de Belalcázar, Lorenzo de Aldana, and Jorge Robledo began to explore the coasts of Peru, crossing both Peru and Ecuador and traveling down the Cauca River. All of them soon became aware of the mineral wealth of the country they had discovered. The Spaniards then began to work the silver deposits in the valleys of the Magdalena, the Cauca and its tributary, the Nechí, and other streams whose riches had been revealed by the first expeditions. As they discovered new lands, they founded settlements on the sites they thought would be future centers for exploitation of mines or other natural resources.

In Colombia, Belalcázar founded the city of Popayán in the upper Cauca Valley in 1536 as well as the towns of Caloto and Santander, in the territory of the Quilichao Indians. Cali was established by Miguel Muñoz under Belalcázar's orders. Jorge Robledo built Cartago in the central part of the valley and with Lorenzo de Aldana founded and settled Anserma in 1539. Santa Fé de Antioquía, in the central Cauca Valley, also was founded by Jorge Robledo. After the settlement had been moved to a spot close to the confluence of the Cauca with the Tomusco River, it quickly became wealthy because of its gold deposits. The city of Arma was built by Miguel Muñoz in 1542, but it soon declined as a result of its unwholesome climate. Río Negro was established in 1545. In the valley of the Las Lanzas River, Andrés López Galarza in 1550 founded Ibagué under the name of San Bonifacio de Ibagué, but he later reestablished it on the site it occupies today. In 1551 Vasco de Guzmán founded Almaguer under the name of Guachicono.

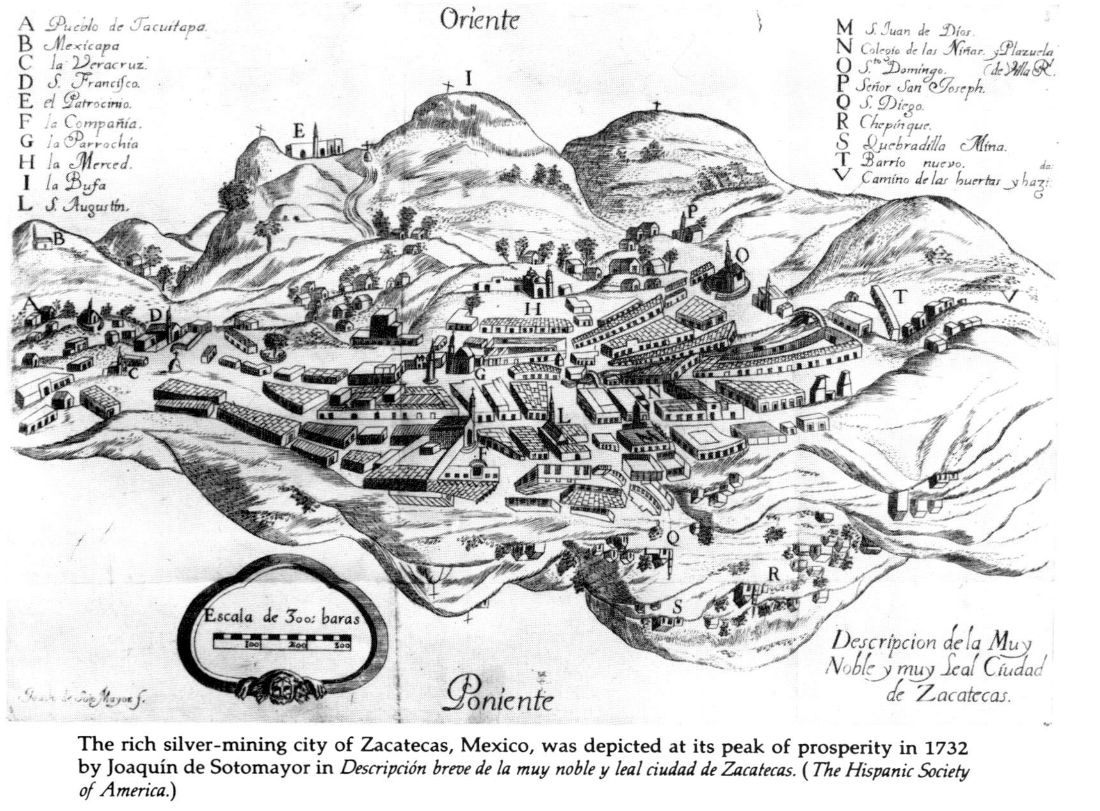

The rich silver-mining city of Zacatecas, Mexico, was depicted at its peak of prosperity in 1732 by Joaquín de Sotomayor in *Descripción breve de la muy noble y leal ciudad de Zacatecas.* (*The Hispanic Society of America.*)

The wealth of La Valenciana at Guanajuato, one of the richest finds in the history of silver mining, stimulated the vigorous growth of all of Mexico in the second half of the eighteenth century. This view of Guanajuato is from C. Nebel, *Voyage pittoresque et archéologique dans la partie la plus intéressante du Mexique* (1836). (*Library of Congress.*)

This charming view of the silver-mining town of Jesús María in northern Mexico was published by John Wodehouse Audubon in his *Illustrated Notes* (1852). The son of the famed American artist-naturalist passed through the village while en route to the California goldfields during the Gold Rush. (*Rare Book Division, The New York Public Library, Astor, Lenox and Tilden Foundations.*)

In their quest for diamonds, the Portuguese went so far as to drain entire riverbeds by diverting the flow through aqueducts. The English traveler John Mawe depicted the scope of such an undertaking on the Jequitinhonha River in Minas Gerais in his *Travels in the Interior of Brazil* (1812). (*The New York Public Library.*)

Remedios was founded by Martínez de Ospina in 1560 and settled chiefly by adventurers who owned Negro slaves. Mariquita, called San Sebastián del Oro during its earliest period, was located on the banks of the Gualí River. Gold was also extracted, though with difficulty, at Guadalajara de Buga, settled in 1559 by Captain Alonso de Fuenmayor under the orders of Don Luis de Guzmán. Gold deposits in Pasto also were exploited around the year 1571.

According to Humboldt, the gold on the west side of the Cauca River is of 12- to 13-carat quality, mixed with silver—the *electrum* of the ancients. The gold of Antioquía is of 19 to 20 carats; of Barbacoas, 21.5; of Indipurdu, 22; and of Chocó, in northwestern Colombia, 20 to 21 carats.

In the district of Chocó, a new metal, platinum, was discovered in its natural state. It appeared in alluvial gold deposits which had been exploited by the aborigines in pre-Columbian times. Don Antonio de Ulloa and Don Jorge Juan, who represented Spain on La Condamine's expedition to Ecuador in 1735 to measure a degree of the meridian, discovered the metal. They published this discovery in their book *Relación histórica del viaje a la América meridional (An Historical Account of the Voyage to South America)* printed in Madrid in 1748, two years before Dr. Brownrigg announced it to the Royal Society in London.

After the middle of the sixteenth century, according to Padre Acosta, alluvial deposits as well as gold in veins were exploited in the cities of Loja and Jaén in the region of Zaruma in the territory of modern Ecuador.[20] Later, these cities were abandoned. In this region the Spaniards also founded the cities of Medina, Sevilla, Logroño, and Zamora. The gold-washing sluices there attracted men from as far south as Lima and as far north as Darién,[21] but battles with the Jívaro Indians forced the abandonment of these cities. In Zamora, gold nuggets weighing three and four pounds each were found, and one, valued at 3,700 pesos, was sent to Philip II. La Condamine in his *Voyage* reported that in 1743 the cities of Loyola and Valladolid had been reduced to two small Indian hamlets. Cuenca, a short distance to the north, was founded in 1557 by Gil Ramirez de Avalos, by order of the Marquis of Cañete, the Viceroy of Peru, near the Indian town of Tumibamba where gold and silver deposits were being worked.

The gold fields of Carabaya, on the Bolivian-Peruvian high plateau, were worked from the beginning and after 1542 were exploited on a large scale. The same is true of those at Oruro, Asillo, and Asangara, where "in past years," says Cieza de León in 1553, "more than 1,700,000 pesos of gold were extracted, so fine that it exceeded the [Crown's] legal standard and this sort of gold can still be found in the river."[22] Cieza adds, "Many found that a single *batea* [a pan used for washing gold] yielded them from 500 to 1,000 pesos, and more than 1,300,000 pesos have been taken out of this river."[23] Sluices more than 6 kilometers (3.5 miles) in length were constructed to wash the gold which was of 23-carat quality. It is said that Charles V was presented with a 100-pound nugget (4 *arrobas*) in the shape of a horse's head. The town of Chaucalla, Peru, was founded in the Condesuyos region in 1550 as a result of gold-mining operations. The Spaniards also exploited the gold in the rivers of the high plateau of Bolivia and Peru, homeland of the Aymará Indians. In Upper Peru, which was to become modern Bolivia, the deposits in the valley of Chuquiabo, or La Paz, were also worked. Of Chayanta and Chilleo, Alvaro Alonso Barba says that they were full of veins of gold with old workings.

The first silver mines to be exploited by the Spaniards in the kingdom of Peru were those of the province of Charcas, now in Bolivia. Among these the first and richest was the Porco mine, already worked by the Incas. This mine was located to the southeast of and not far from Potosí. Its ores are chiefly pyrargyrite, or ruby silver, which the Peruvians called *cochizo,* or *rosicler.* One of the mines in Porco was exploited by Hernando Pizarro about 1549. Cieza de León, who was still in Peru at that time, says that Pizarro "received annually from the pure metal that was extracted more than 200,000 gold pesos." And Cieza adds, "Formerly Porco was a very rich place, and so it is still, and it is thought that it will always be so." Following the fortunes of the struggles among the conquistadors, possession of these mines oscillated among the victors in their many and bloody factional struggles until at last Presidente Pedro de La Gasca placed them under control of the Crown.

The famous mines of the Cerro de Potosí were discovered in 1545 by an Indian named Guallpa, in the hollow left by an uprooted shrub. Guallpa tried to keep the secret, but it was uncovered by an Indian from Jauja named Guanca. Seeing that it was going to be difficult to work the vein, Guanca told his master, a Spaniard named Villaroel, about the discovery, and Villaroel registered the mine in his own name and Guanca's on April 21, 1545, in the settlement of Porco.[24] "And because the Indians call the mountains and high places *potosí,* " Cieza de León tells us, "it acquired the name of Potosí. . . ."

In the rich lode Guanca discovered, "The metal," according to Padre Acosta:

> was an outthrust in the form of crags raised above the surface, like a crest three hundred feet long and thirteen wide. . . . The color of this mountain (crowned with porphyry) is a sort of dark red; it is very graceful to look upon, like a tent in the shape of a sugarloaf. It rises above and dominates all the other mountains that are around it. The climb is difficult even when it is all made on horseback; it terminates in a round peak; the perimeter of its base is about a league [three miles]; from the peak of this mountain to its foot and base, there are 1,624 *varas* [yards] of common measure. . . . On this mountain, at the bottom of its slope, is another small one that rises from it, which formerly had some mines of divers metals that were found in little clumps, and not in a fixed lode, and were very rich although few in number. It is called 'Guainapotosí,' which means 'Potosí the Younger.'[25]

The lodes were so rich that in the one discovered by Guanca, Zárate tells us, "wherever they assayed they took out pure or almost pure silver, and the least that was extracted was eighty marks per *quintal,* which is the greatest riches that have been seen or read of in any mine ever worked."

Pedro Cieza de León says that six years after the mines of Potosí were discovered, the "Royal Fifth" (the share deducted for the Crown) alone amounted to "more than the Spaniards had taken from Atabaliba [Atahualpa] and more than they found in the city of Cuzco when they discovered it."[26] Cieza also tells us that:

> although at this time Gonzalo Pizarro was in rebellion against the Viceroy, and the Kingdom full of unrest caused by this rebellion, the slopes of this mountain were settled and many large houses were built, and the Spaniards made their chief settlement in this region, bringing the courts of law to it so that the capital was almost deserted and abandoned; and thus mines were opened, and high up on the mountain five very rich lodes were discovered, which are called Centeno's, Rich Lode, Tin Lode, and the fourth is called Mendieta's, and the fifth Oñate's. These are lodes that run from north to south, and each one is exploited by several mines.[27]

Exploitation of the mines of Potosí, the rapid increase in the region's population, the founding of the city, its life and vicissitudes, and the universal fame it acquired within a very short time constitute one of the most fantastic among the many stories of the Indies.[28] The settlement went forward with great rapidity from the very moment of the mine's discovery in 1545. The official founding of the city dates from 1561 when a royal cedula, issued in Valladolid by the Princess in the name of Philip II, "endorsed the Act executed in the City of the Kings [Lima] on 7 November 1561, to grant it the title of *Villa* [a city enjoying royal privileges], and [decreed] that it is named and shall be called the Imperial City of Potosí."[29] About 1570, twenty-five years after its founding, the number of inhabitants had risen to more than 120,000 souls, coming from every corner of America, the Iberian Peninsula, and many Mediterranean countries. According to a census taken in 1611, at that date "the inhabitants of the City numbered 150,000 living within a radius of two leagues around, including very noble families, Indians, and Spaniards of both sexes."[30]

Life in Potosí was luxurious and splendid. The city was proud of its escutcheon, granted by Charles V, and of its motto "Most Faithful and Noble City" bestowed on it by the Viceroy Don Francisco de Toledo. And thus the eighteenth-century historian Cañete tells us:

> The Mayors have always been very lavish, and spend excessively during the year in which they hold office; for apart from the opulent tables they set, the refreshments and entertainments they provide at their receptions, each one brings in two, three, and even four pages, dressed in rich cloth and covered with braid, who have the title of Ministers. . . . This town was in olden days so opulent and rich, and so splendid in its official ceremonies, that at the time of the coronation [sic] of His Most Serene Emperor Charles V eight millions were spent; in the ceremonies for Our Lord Don Philip II, six millions; and likewise with the other Kings of Castile. All this came from the citizens of the town, for each contributed according to his means through a quota administered by the magistrates. The same is also proved by the exaggerated sums with which many gentlemen dowered their daughters. General Pereyra married off his daughter Doña Plácida Eustaquia in 1579, giving her a dowry of 2,300,000 pesos; General Mejía dowered his daughter with 1,000,000 in 1612; Doña Catalina Argandona took a dowry of 800,000 pesos with her, in addi-

> tion to some vineyard properties, when she married Don Luis de Esquivel. Doña Ursula Obando had the same amount in 1629; and, finally, up to the year 1647, they tell of more than eight dowries, the smallest of which was more than 200,000 pesos; on the other hand, nowadays we find not a single dowry of more than 50,000 pesos free and clear.[31]

Attracted by so much wealth, a market had arisen which was much greater than all others in Peru. According to Cieza de León, it sometimes sold goods worth "40,000 gold pesos in one day, and I believe that there is not a single fair in the world that can equal the transactions in this market."[32]

The historian Luis Diez del Corral was struck by the variety of precious objects which arrived in Potosí from all over the world. Beside European merchandise, there were silks and carpets from Africa; perfumes from Arabia; diamonds from Ceylon; crystal, ivory, and precious stones from East India; and spices, aromatics, and porcelain from Ternate, Malacca, and Goa.

In addition to luxuries, the Potosí market also stocked the staples vital for the food supply of a city located in the midst of an inhospitable desert.[33] The modern historians Lewis Hanke and Gunnar Mendoza say in respect to this:

> All aspects of life, even religion, were directly affected by the flow of silver that descended from the Mountain. The miners spent ostentatiously not only during their lives, making showy donations to churches and monasteries, but also when they died, arranging for sumptuous funeral ceremonies. Sports were popular and offered large profits to the impresarios of jai alai, to which Viceroy Toledo was opposed because it brought together crowds of idlers and wasted time that would have been better employed in mining operations. No one could escape the desire to produce and take advantage of the opportunities for profit offered by Potosí.[34]

The fame of Potosí's wealth quickly spread throughout the world and resulted in the fact that in every language the phrase "worth a Potosí" was coined as a synonym for wealth. The expression current in English literature is "as rich as Potosí."

The city's economic and social life naturally followed the ups and downs of mining operations in the mountain. It was linked not only to discoveries of new lodes, but also to techniques for their exploitation, of which we will speak later. By the end of the eighteenth century, however, the riches of the Potosí mines gradually diminished as the shafts were deepened and silver sulfides and chlorides began to predominate over the native silver. Within the next twenty-five years, the decline was plainly visible and caused serious concern for the future. Neither are mines limitless in ore nor do they ever renew themselves. The Mountain of Potosí reminded Antonio Ulloa who saw it in the eighteenth century of a beehive because of its innumerable mouths and the concavities and mines in its interior. In a short time, the mountain was doomed to be mined out.

A visit to the Imperial City today—"that tremendous economic and social phenomenon which developed within and on the slopes of that Cerro Rico," as Don Armando Alba the contemporary historian and native of Potosí has said—makes a profound impression. At an altitude which is the natural habitat of condors and llamas rather than of men, it is surrounded by an almost sterile landscape. The mountain itself is now reduced to the exploitation of tin ore in mines more than 3,000 feet deep. The visitor is immediately impressed by the historic atmosphere of the place and feels attracted by the culture and good manners of the inhabitants. He can observe in its churches and fine buildings, among them the peerless mint, the memories the city retains of its splendid historical past.

Questions have often been asked concerning the mining and working in America of metals other than gold and silver. Though known and necessary, other ores such as iron, lead, and copper were not exploited for easily explainable economic reasons. Alexander von Humboldt explains the phenomenon in a long and sophisticated piece of economic analysis:

> The mountains of the new continent, like those of the old, contain iron, copper, lead, and a great many other mineral substances indispensable to the requirements of agriculture and the arts. If in America man's labor has been dedicated almost exclusively to the extraction of gold and silver, it is because the members of a society act out of considerations very different from those which must influence society as a whole. In every place where the soil can produce both indigo and maize, cultivation of the former prevails over that of the latter, even though it would be more in the public interest to prefer the plants which serve as human nourishment over those which furnish a means of foreign exchange. In the same

way, on the slopes of the mountain ranges, iron or lead mines lie abandoned no matter how rich they may be, for the settlers' attention is entirely fixed on the gold and silver lodes, even when their outcroppings present only meager indications of riches. So great is the attraction of those precious metals that by general agreement they have become the representative signs of subsistence and labor.

The Mexican people are no doubt able to procure, through foreign trade, everything that is not supplied to the country they inhabit. But, in the midst of their wealth in gold and silver, they suffer each time there is an interruption in trade with Spain or the rest of Europe, each time a war interferes with communication. Sometimes twenty-five to thirty million pesos pile up in Mexico, while factories and mining work are hampered by a shortage of steel, iron, and mercury. A few years before my arrival in New Spain, the price of iron had risen from 20 francs the quintal to 240; that of steel, from 80 to 1,300. During these periods of total stagnation of foreign trade, Mexican industry momentarily revives. It is then that production of steel is undertaken, that the iron and mercury ores which lie hidden in the mountains of America begin to be used. It is then that the nation enlightened about its own interests realizes that its true wealth consists in the abundance of consumer products, of *things,* and not in the accumulation of a *sign* that represents them. During the next-to-last war between Spain and England, there was an attempt to exploit the iron mines in Tecalitlán, near Colima, in the intendancy of Guadalajara. The Mining Tribunal spent more than 150,000 francs to extract the mercury from the lodes of San Juan de la Chica. But the effects of this laudable effort were of short duration. The Peace of Amiens put an end to the enterprise, which seemed to give to mining operations a more useful direction for public prosperity. Scarcely had sea traffic been restored when iron, steel, and mercury were again being sought by preference in European markets.

As the population of Mexico increases and its inhabitants come to depend less upon Europe, they will begin to fix their attention on the great variety of useful products to be found in the bosom of the earth, and the system of mining exploitation will change. An enlightened administration will encourage projects intended for the extraction of mineral products of *intrinsic value.* Private persons will no longer sacrifice their own interests, and those of the public welfare, to stubborn prejudice; they will realize that exploitation of a coal, iron, or lead mine can become as profitable as that of a seam of silver. In the present state of Mexico, precious metals are virtually the settlers' only industry, and when in this chapter we employ the word 'mine' *(real, real de minas)* it will be understood, unless expressly stated to the contrary, that we are referring to a gold or silver mine.[35]

Notes

1. Padre Bartolomé de Las Casas, *Historia de las Indias,* book I, chaps. LXV–LXVI. My account of mining operations follows along the general lines of the important book by Modesto Bargalló, *La minería y la metalurgia en la América Española durante la época colonial* (Mexico, 1955).

2. Andrés de Tapía, "Relación," in *Crónicas de la Conquista de México* (1950).

3. Bernal Díaz, *Verdadera historia de la conquista de Nueva España,* chap. CXXXVII.

4. Modesto Bargalló, *La química inorgánica y el beneficio de los metales en el México prehispanico y colonial* (Mexico City, 1966).

5. Hernán Cortés, *Fourth Letter,* Oct. 15, 1524.

6. *Archivo General de Indias.* Manuscript, modern copy by F. del Paso y Troncoso, in the National Museum of Mexico.

7. For Borda's biography, see Manuel Toussaint, *D. José de la Borda restituido a España* (Mexico, 1933).

8. For Borda's later career, see David A. Brading, "Mexican Silver Mining in the Eighteenth Century: The Revival of Zacatecas," *Hispanic American Historical Review,* vol. L, no. 4 (November 1970), pp. 665–81.

9. Francisco X. de Gamboa, *Comentarios a las ordenanzas de minas* (Madrid, 1761), p. 67, n. 1.

10. Ibid.

11. Alexander von Humboldt, *Essai politique sur le Royaume de la Nouvelle-Espagne,* book IV, chap. 11.

12. Domingo Lázaro de Arregui, *Descripción de Nueva Galicia* (1629), F. Chavalier, ed. (Seville: Consejo Superior de Investigaciones Científicas, 1946).

13. *Testimonio de Zacatecas* (1718) (Mexico: Imprenta Universitaria, 1946).

14. Humboldt, op. cit., chap. 7.

15. Pablo Herrera Carrillo, quoted by Gustavo P. Serrano, *La minería, y su influencia en el progreso y desarrollo de México* (Mexico, 1951).

16. By 1968 the company was approaching an annual production of one million tons, and the city has become an industrial center with a population of about a million and a half.

17. Humboldt, op. cit.

18. Bargalló, *La minería,* chaps. 6, 19, and 24.

19. Alvar Núñez Cabeza de Vaca, *Naufragio y comentarios* (Madrid: Espasa-Calpe, 1936).

20. Acosta, *Historia natural,* book IV, chap. VI.

21. Vidal de la Blanche, *Geografía universal,* pp. 436–437.

22. *Guerras civiles del Perú,* chap. XI.

23. *La Crónica del Perú,* chap. CII, cited by Luis Díez del Corral, *Del Nuevo al Viejo Mundo* (Madrid: Revista de Occidente, 1963).

24. Acosta, op. cit.

25. Ibid.

26. *La Crónica del Perú,* chap. CIX.

27. Ibid.

28. Extremely detailed information on the vicissitudes of life in Potosí, statistical data on production, and information on administration and government appear in innumerable documents, mostly unpublished, in the archives of Potosí, Sucre, Lima, and Buenos Aires and in the *Archivo General de Indias* in Seville. Among the chronicles and histories, those of most outstanding importance are the *Relación general del asiento de la Villa Imperial de Potosí* by Luis Capoche, written at the end of the sixteenth century and not published until 1959; the *Historia de la Villa Imperial de Potosí* (Providence, R.I.: Brown University Press, 1965) by Bartolomé Arzáns de Orsúa y Vela, which we have already mentioned, written in the early years of the eighteenth century and very recently published; and the *Guía histórica, geográfica, física, política, civil y legal del Gobierno e Intendencia de la Provincia de Potosí* by Pedro Vicente Cañete y Domínguez, written in 1787 and reprinted in 1952 by the Editorial Potosí in Potosí.

29. Cañete y Domínguez, op. cit.

30. Ibid. See also Lewis Hanke's preliminary study in the Biblioteca de Autores Españoles edition (Madrid, 1959) of Capoche's *Relación general del asiento de la Villa Imperial de Potosí* (1581), as well as this important work itself.

31. Cañete y Domínguez, op. cit.

32. *La Crónica del Perú,* chap. CX.

33. *Del Nuevo al Viejo Mundo,* pp. 72–73, citing Arzáns de Orsúa y Vela, op. cit.

34. In their prologue-biography to Bartolomé Arzáns de Orsúa y Vela, op. cit.

35. Humboldt, op. cit., chap. 9.

V

The Special Case of Brazil

Brazil, the portion of the New World reserved for Portugal when these lands were divided between Spain and Portugal by Alexander VI in his Bull of 1493 and the Luso-Spanish Treaty of Tordesillas of 1494, represents a special situation in Latin American history. In contrast to Spain's possessions, discoveries of precious metals in the lands of Brazil were scanty and long delayed. In fact, no appreciable amounts of gold and silver were found for two centuries following its discovery and conquest.

Brazil's northern coasts were first sighted and explored by Spanish expeditions led by Vincente Yáñez Pinzon in 1499 and Alonso de Ojeda, accompanied by Amerigo Vespucci, in 1500. News of the discovery made only a slight impression in Spain and certainly gave little hint of the huge continent to the south. No mineral riches were discovered, and in any event, the eastern "bulge" of South America belonged to Portugal.

Then, early in 1500, Pedro Alvares Cabral, while leading a Portuguese fleet to India, sailed farther west than planned and, on April 22, found a land he called the island of Santa Cruz.[1] The ship that brought the news back to Lisbon opened a new chapter in history for Portugal and the American continent, for Cabral had discovered not an island but the mainland of Brazil. That fact was ascertained by Vespucci the next year on an expedition commissioned by the Portuguese Crown to explore Cabral's landfall. The expedition also brought back samples of brazilwood from which a red dyestuff is derived; the region in fact received its name from the tree. By 1503 a Portuguese *converso,* Fernão de Noronha, received a concession from the Crown to open *feitorias* (trading posts) and to cut brazilwood for export. A demand for dyestuffs in Flanders, France, and England, which were engaged in intense competition with the Italian textile industry, assured a profitable market. At the same time, however, the opening of the new trade route to India via the Cape of Good Hope focused the energies of Portugal's military and merchants upon the conquest and economic exploitation of that large and rich section of the globe. As a result, Noronha was followed by few of his countrymen. In fact, a large proportion of the early settlers were either *degredados,* petty castaways, or converted Jews fleeing the Inquisition. French and English ships visited Brazil to cut brazilwood, and warfare broke out sporadically. Each nation raided the other's trading posts and attacked merchant vessels or even fleets. These actions left the future of Brazil in the balance.

Portugal ultimately won the struggle for Brazil because the Portuguese introduced agriculture as an adjunct to the mercantile operations of the feitorias. In addition to acclimating Eastern Hemisphere crops to Brazilian conditions, the Portuguese learned how to raise native crops, particularly corn and manioc. They also began to recognize the possibility of growing sugar cane, one of the most profitable crops of the time and one with which the Portuguese were familiar as a result of their operations in the Madeira and Cape Verde Islands. Thus, Portugal developed a great long-run advantage: a viable economic base to support and feed a growing population of European settlers. Martim Afonso de Souza drove out the French traders and merchants in 1530–1533, and before he returned to Portugal, the King set up a new system of government and land donations. These were based on the Azores and Madeira models and in some ways resembled the later British proprietary colonies in North America. Brazil was now assured of remaining a Portuguese domain.

For about a hundred years Brazil expanded economically and geographically. As Portugal's power waned in the East, the flow of men and capital to Brazil increased. Negro slavery became a major institution. Sugar, tobacco, cotton, citrus fruits, hides, and brazilwood provided the base for a wealthy planter aristocracy. Portuguese capital built large sugar mills, and Bahia and Pernambuco became important cities.

Despite the apparent lack of minerals, however, foreigners still coveted Portugal's colony. A French attempt to set up a settlement at the site of Rio de Janeiro in 1550 failed and the invaders were driven off in 1567. The Dutch later made a more determined effort, seizing a large section of the coast, establishing a capital at Pernambuco, and holding a major portion of the sugar lands from 1630 to 1654.[2] The ultimate expulsion of the Dutch was almost exclusively a Brazilian enterprise, and it contributed to incipient feelings of nationalism, particularly since the three main leaders of the resistance were a white man, an Indian, and a Negro.

By the middle of the seventeenth century, however, the world economic picture had changed and the prosperity of Brazil began to decline. By a supreme effort, Portugal had freed itself from the "Babylonian Captivity" of Spain in 1640, but the struggle to maintain its independence and carry on war against its larger neighbor on the Peninsula and the Netherlands on the high seas was extremely costly. Although the Dutch were expelled in the 1650s, they had learned in Pernambuco the technique of growing sugar. Meanwhile Portuguese capital was in short supply and the markets of Europe were reduced since England, France, and Holland had seized "sugar islands" in the Caribbean from the Spanish and were able to supply their home markets as well as sell their colonies' surplus abroad. Yet despite all this, trade between Brazil and Europe was by no means at a standstill. The English explorer and buccaneer William Dampier who visited Bahia in 1699 reported:

> A great many merchants always reside at Bahia; for 'tis a place of great trade. I found here above 30 great Ships from Europe, with 2 of the King of Portugal's Ships of War for their Convoy besides 2 Ships that traded to Africa, either to Angola, Gambia or other Places on the coast of Guinea and abundance of small craft that only run to and fro on this Coast, carrying commodities from one port of Brazil to another.[3]

The contrast between the pattern of Portuguese settlement in Brazil and the Spaniards' in other parts of the New World is interesting. While the lure of mines expanded the dominions of Castille into the deserts of northern Mexico and the thin air of the high Andes, commercial agriculture kept most Portuguese along the seaboard close to the lines of transatlantic communication. Conquering the Indians of the interior in order to seize the land was not easy, and except for slave-hunting expeditions, the natives were left alone. These expeditions, called *entradas* or *bandeiras,* meaning companies of backlands militia men, covered much of the interior of Brazil in search of Indians and anything else of value, including gold and precious stones.[4] Their most profitable raids were on the Paraguayan missions run by the Jesuit fathers. Missions provided convenient undefended concentrations of docile natives to meet the labor demands of the Brazilian coast. While blacks were preferred and sold at a premium over Indian slaves, planters were often willing to settle for a cheaper Indian labor because of interruptions in convoys and the high price of Africans. The work of the *bandeirantes* (members of the bandeiras) laid open much of the interior of Brazil, but it did not establish permanent centers in the backlands to support a viable economy.

Brazil's social and economic organization also militated against the establishment of a strong, centralized colonial government. Land had been distributed in large blocks, and the recipients were given virtually feudal privileges. Consequently, the country tended to be a collection of loosely connected nuclei scattered along the coast. These separate settlements did not maintain close contacts except when faced with great problems such as the Dutch invasion. A Captain-General, whose title emphasizes his major function as military, resided at Bahia and carried out the orders of the Overseas Council in Portugal, but a permanent Viceroyalty was not established until 1720. The economic pattern tended to tie the various settlements to Lisbon since they produced nearly the same products for export and needed similar imports. While juridical, ecclesiastical, and cultural institutions were established in Brazil, the mother country encouraged dependence. Young men desiring higher education had to cross the Atlantic to Coimbra, and no printing press was definitely established in Brazil until the nineteenth century!

Compared to the capitals of the Spanish Viceroyalties with their carefully planned towns and their universities, print shops, and mints, the cities of Brazil seemed provincial. Yet the plantation owners and business men contributed to the building of handsome churches and monasteries, and some of the settlers were well educated and patrons of the arts. In 1612, for instance, Lope de Vega dedicated twelve of his plays to *"Duarte de Albuquerque Coelho, Capitão & Gouernador de Pernambuco, na Noua Lusitania."* The clergy meanwhile felt that some of the wealthier families were overly ostentatious; the missionary Father Fernão Cardim, writing of the Pernambucanos, commented:

> The men go clad in velvet, silks, and damask and spend their money freely on valuable horses with saddles and other trappings of the same materials. The ladies as well make a show of luxury and are fonder of gay parties than they are of devotions. . . . One encounters more vain display here than one does in Lisbon.[5]

Another Jesuit, Frei Vicente do Salvador, in his *História do Brasil* (1627), was less hostile to riches and lamented the fact that although only an invisible line down the spine of the Andes divided Peru from Brazil, it was the Spaniards who discovered so many mines while the Portuguese had none. Not so, said Antônio Vieira, "the greatest preacher ever to use the Portuguese language," who asked in a sermon in the mother church of Belem do Pará in 1656:

> What advantages have come to Spain from her famous Potosí and other mines in this same America? Spain herself confesses and deplores that the service they did her was nothing more than to depopulate and impoverish her. They dig and transport the silver and foreigners acquire and enjoy it. The others benefit from the substance of the precious metals while they are left with the slag . . . [and, he adds,] none of this is for the relief or benefit of the people.[6]

It was the discovery of gold, however, which confirms the thesis of this study: that without mining centers to produce great amounts of wealth and attract large pioneer populations, the settlement of new areas is extremely difficult. Certainly the undeveloped state of the interior of Brazil can be so explained.

Early attempts to find precious metals were not too successful in Portuguese America. Some alluvial gold was discovered as early as 1541, and Dutch goldsmiths were active in connection with the Schetz enterprises in Santos soon after.[7] Toribio de Ortiguera explored the Maranhão River, the headwaters of the Amazon, around the year 1561 and left an account saying that he had seen gold objects possessed by the natives, but could not find the gold's source.[8] By 1572 the alluvial, gold-producing streams of Paranagua had been discovered, but production there was never large. One of the most knowledgeable Brazilian settlers, Gabriel Soares de Sousa, who arrived in 1569 and became very wealthy, owning a sugar plantation as well as houses and property in São Salvador (Bahia), was convinced that he could acquire even greater riches in the hinterland. He based his belief on samples of gold, silver, and precious stones which his brother João Coelho de Sousa had collected in the *sertão* before his death. Gabriel Soares therefore went to the Court of Madrid in order to obtain all possible rights of discovery. It was in connection with this request that he wrote the *Tratado descritivo do Brasil em 1587,*[9] the most accurate account of Portuguese America made during the sixteenth century. In December 1590 he finally obtained the desired document which named him Capitán Mayor y Gobernador de la Conquista y Minas del Rio São Francisco with privileges of a truly feudal character. More than 300 people accompanied the wealthy planter and explorer across the Atlantic in 1591, and they soon headed for the interior. Sickness, however, was the undoing of the entrada, and Gabriel Soares himself died at the headwaters of the Paraguaçu River before the expedition had progressed very far. If this attempt had succeeded, the history of Brazil might have been very different.

More practical metals also were exploited to some extent. An iron smelter was set up at Araçoiaba in Sorocaba near São Paulo by 1590 and another at Ubatá in 1603. Both functioned until 1629 when an edict from the Metropolis in Portugal put an end to their activity; the mother country did not want competition from the turbulent Paulistas even though these enterprises were never commercially significant. During the final years of Spanish rule over Portugal (1580–1640) Pedro Teixeira, commanding an expedition of 70 soldiers and 1,200 Indians in 60 canoes, explored the Amazon from its mouth as far as the Napo River in Ecuador and then marched overland. On August 15, 1638, he reached the city of Quito where he was warmly welcomed. Teixeira also explored the region around the Yupur River, another tributary of the Amazon, and reported finding gold jewelry among the natives there, but no formal prospecting work was begun.[10] In the late seventeenth century the Portuguese Crown sent Don Rodrigo de Castelo-Branco, a Spaniard well-versed in mining, to explore its dominions in the New World. Experienced in prospecting in Upper Peru and Potosí, he still found nothing in Brazil.[11]

The turning point came during the years from 1693 to 1695 when bandeirantes at last found rich gold placers in the small Velhas River in what is now the state of Minas Gerais.[12] Within a decade the relatively placid, bucolic life of Brazil was shattered forever. Not only was Brazil's economy reoriented, but its social structure and political life also were drastically altered in the face of the outpouring of enormous wealth. The disruption was so complete

A rare 1584 watercolor by an unknown artist depicts the great mountain at Potosí, with trains of llamas transporting silver ore down the slopes to the refineries that were built for processing the precious metal. (*The Hispanic Society of America.*)

A somewhat fanciful European interpretation of the workings within the world-famous silver lode at Potosí appeared in Théodore de Bry's *Americae* (1602). (*Rare Book Division, The New York Public Library, Astor, Lenox and Tilden Foundations.*)

Huancavelica, Peru, was the site of quicksilver mines that were first exploited by the Portuguese Enrique Garcés in the 1560s. Following Bartolomé de Medina's discovery of the so-called patio method in 1554, quicksilver became essential in the production of silver. The view of Huancavelica is from Mariano Eduardo de Rivero y Ustariz, *Colleción de memorias científicas, agrícolas, e industriales* (1857). (*Library of Congress.*)

At the Hacienda de Proaño in Zacatecas, Mexico, silver ore from the Fresnillo mines is being amalgamated by the patio process. In this oil painting by Gualdi, horses are being used to tread the ground and moistened masses of ore, or *repasos*. (*Museo Nacional de Historia de Chapultepec.*)

that the royal authorities simply could not maintain order against the onslaught of the avalanche of wealth and the people seeking it. In a cry of despair over the revolutionary changes this money brought, Father Apolinario de Conceição, O.F.M., lamented, *"A meu Deus, com ouro nos quereis castigar!"* (My God, with gold you wish to punish us!) The Portuguese government was similarly agitated by the prospect of the jealousy and cupidity of foreign nations being provoked anew by the exceedingly rich mines. Faced with one of the largest gold strikes in history, the authorities found themselves in constant perplexity as they attempted to keep up with the changes the mineral wealth had brought and tried to contain its effects.

Padre Antonil's *Cultura e opulência do Brasil,* which appeared at Lisbon in 1711, describes conditions during the period:

> They say that the first discoverer was a mulatto who had been at the mines of Paranguá and Curitiba. He was going with some Paulistas into the sertão to seek Indians, and on reaching Mount Tripué went down to the stream, [near the town] which is known today as Ouro Prêto, with a wooden bowl to obtain water. After scraping it along the bank he saw that it contained nuggets the color of steel without knowing what they were.[13]

These nuggets were sent to the Governor of Rio de Janeiro, Artur de Sá:

> And when they were examined they turned out to be the purest gold.
>
> Half a league from the Ouro Prêto riverbank another mine was found which is called that of the Ribeiro de Antônio Dias; and another half league from there, the mine of Ribeiro do Padre João de Faria; and near this, a little more than a league away, the mines of Ribeiro de Bueno and Bento Rodrigues . . . and all these mines took the names of their discoverers, who were all Paulistas [bandeirantes].

Although few exact figures exist, we know that migration into the gold fields was enormous. Portugal, in fact, feared that the coast would be depopulated. In all, an estimated 300,000 Portuguese emigrated to the colony in the eighteenth century, with most of them heading for the gold fields. Over the major highways from São Paulo, Rio de Janeiro, and Bahia, a stream of pioneers and adventurers poured into Minas Gerais. Antonil reported:

> It would be difficult to calculate the number of persons who are now there. However, those who have stayed there for a long period during these last years and have visited [all the mines] say that more than thirty thousand souls are employed, some in prospecting, others in directing the prospecting of the gold washings, and others in doing business, selling and buying the essentials not only for life but for pleasure, with greater activity than in the seaports.[14]

There was a King's Superintendent of Mines, a Crown Procurator, and a Crown Representative *(guarda mór)* in the region, but the smelting houses were at Taubaté (São Paulo) and Parati in the state of Rio de Janeiro. As a result, collection of the Royal Fifth was extremely difficult. Again there was great rivalry between the Paulistas and the *Emboabas,* or newcomers, from the coast and overseas. However, the prospectors did set up rules concerning the allotments or distribution of the mines, and Antonil is again our best source:

> To avoid the confusion, tumult, and deaths that would arise from the discovery of the gold washings the following method of allotment was agreed upon. The discoverer has the first claim as discoverer and a second as a miner. Next follows the King's share and then that of his representative. The rest are distributed by lot. Those known as whole claims are thirty square *braças* [sixty-six by sixty-six meters], and such are those of the King, the discoverer, and the royal representative. The others which are drawn by lot have an area proportionate to the number of slaves brought for prospecting with two square *braças* allowed for each slave or Indian so used. Thus a man with fifteen slaves receives a whole claim of thirty square *braças.* To be admitted to the distribution by lot one must submit a petition to the superintendent of these allotments, who receives an *oitava* [one-eighth of an ounce] of gold for filing the petition, as does his notary. And it sometimes happens that five hundred petitions are presented, which means that the superintendent and the clerk take in one thousand *oitavas.* And if, because some claims fail to produce, all the miners do not extract from them an amount equal to what they paid, they therefore look for other claims as soon as new washings are discovered. The King's claim is sold straightway to the highest bidder. Anyone can sell or exchange his claim, and this has given and continues to give rise to many different arrangements, for one miner may extract much gold from a few *braças* while others extract little from many *braças.* There was one man who sold a claim for more than a thousand *oitavas*

> from which the buyer extracted seven arrôbas [an arrôba is 15 kilograms] of gold. This goes to prove that whether or not gold is found in a claim is simply a turn of good or bad fortune.[15]

For a few years the Portuguese government continued to regard its mineral wealth as but a passing thing, feeling that the *real* gold mines of Brazil were to be found in the plantations of sugar and tobacco. Faced with a threatened depopulation of the coast and possible foreign invasion and conquest of Brazil, Lisbon at first enacted laws both to inhibit migration to the interior and to limit its economic expansion. Roads were closed, numerous permits were required, foreigners were expelled, newly imported slaves were distributed in a manner discriminatory to the mining regions, and royal officials were admonished time and again not only for laxity in enforcing the regulations but for being parties to their violation for personal profit. Even the clergy were not exempt; a number of them who entered the mining districts neglected their calling and engaged in trade, particularly smuggling, since they were not subject to search.[16]

Finally, at the end of the first decade of the eighteenth century, the Overseas Council in Lisbon realized that emigration to the interior could not be regulated by law and it was best to accept the situation. In 1710 a new Captaincy-General of São Paulo and Minas Gerais was established under the governorship of Antônio de Albuquerque, an able, tactful, and forceful individual who brought order to the region. It was he who elevated the town of São Paulo to a city and the mining camps of Ribeirão do Carmo, Ouro Prêto, and Sabará to townships. In conjunction with the miners he also established the annual tax of the Royal Fifth at from 8 to 10 oitavas per bateia (bowl in which gold was washed). The wealth of Minas Gerais was now the main sustenance of the economy of Brazil and a major prop to the economy of Portugal, more than making up for the slump in the sugar market. The Council severed Minas Gerais from the province of São Paulo in 1720 and set it up as an independent *capitania* of equal status. It removed the labor and migration restrictions, reopened all roads to the interior, and began improvements. The gold boom was now in high gear.

The leading town in the mountainous regions of the interior had the picturesque name of Villa Rica de Nossa Senhora de Pilar de Ouro Prêto (Rich City of Black Gold of Our Lady of the Pillar). The place nearby where diamonds were later discovered in the 1720s, although officially named the *arraial* (encampment) of Tijuco, was generally referred to as the *Distrito Diamantino.* Tijuco was legally given the name Diamantina only in 1831. Indeed there were so many mining centers in the backlands of eastern Brazil in the early eighteenth century that the high plateau region surrounding the Serra do Espinhaço came to be known as Minas Gerais, General Mines.

Antonil graphically described the variety of settlers drawn to the sertão at the beginning of the eighteenth century:

> Each year great numbers of Portuguese and foreigners come in the fleets, headed for the mines. From the cities, towns, plantations and hinterland of Brazil there go whites, browns, blacks and many Amer-Indians who are in the service of the Paulistas. The mixture is of every kind and condition of persons; men and women; young and old; poor and rich; nobles and plebians; laymen, priests and monks of different orders, many of whom do not have either convent or house in Brazil.[17]

Villa Rica de Ouro Prêto soon became the most thickly populated community of the area. Less than three decades after its founding, a book was published in Lisbon entitled *Triumph of the Eucharist, [an] Example of Lusitanian Christianity . . . in Villa Rica, Court of the Captaincy of the Mines . . . Dedicated to the Sovereign Lady of the Rosary by the Black Brothers of Her Brotherhood, and at Their Request Brought to Public Notice by Simam Ferreira Machado, [a] Native of Lisbon, and Resident in the Mines* (Lisbon, 1734). The author of this work states:

> In this town live the chief merchants, whose trade and importance incomparably exceed the most thriving of the leading merchants of Portugal. Hither, as to a port, are directed and collected in the Royal Mint the grandiose amounts of gold from all the Mines. Here dwell the best educated men, both lay and ecclesiastic. Here is the seal of all the nobility and the strength of the military. It is, by virtue of its natural position, the head of the whole of America; and by the wealth of its riches it is the precious pearl of Brazil.[18]

In the year 1721 new gold strikes were made at Cuiabá in the present state of Mato Grosso, followed five years later by discoveries in the present state of Goiás. The amount of gold mined in Goiás soon rivaled the output of Cuiabá. If prices of foodstuffs and supplies were high in Minas Gerais, in the new areas they were astronomical, and commercial pursuits, farming, and livestock increased quickly to become at least as profitable as mining. Finally, the agricultural life remained and gave the region a viable economy after the mines were exhausted. Still other mines along the Sararé River started a new rush, but the cost of supplies was so high that hunger was commonplace. However, gold seekers continued to pour in, including people who had been disappointed in the older areas as well as new arrivals from the coast and Portugal. By 1737, Goiás was a provincial seat and twelve years later became a Capitania. By then it was so prosperous that the miners were reported ordering the finest silks from abroad. The opening of these interior gold fields led to a reappraisal of the supply routes, and soon a road ran to Goiás along which cattle and foodstuffs started to move in abundance. Soon after, bandeirante explorations showed that goods brought into the Mato Grosso and Goiás markets via the Amazon tributaries from warehouses in Belém could compete with those brought in on the overland routes from Rio de Janeiro and São Paulo.

An event of profound importance in the history of Brazil was the discovery, referred to above, of the Diamond District of Minas Gerais.[19] In the mid-1720s a faiscador by the name of Bernardo da Fonseca found diamonds among the gold washings of a place called Morrinhos in the Sêrro do Frio region but disclosed the information only to a few friends. Other prospectors had also picked up some of the stones but believed that they were crystals without value. It is said that they were even used to keep score in card games! Under pressure from a judge, Antônio Ferreira do Vale, who had obtained some of the diamonds, Fonseca was forced to sell his site. Consequently in 1727 he made a trip to Villa Rica to inform Governor Lourenço de Almeida of the whole story. The latter waited two years before reporting the news to Portugal and then did only after learning that Padre Antônio Xavier de Sousa was sailing for Lisbon and would inform the King of the discovery. The Governor explained that he could not tell whether "the little white stone" had any value—opinions differed in the district—but that jewelers in Lisbon would know. Meanwhile he suspended gold washing and mining in the Sêrro do Frio and canceled all allotments that had been granted. Since Dom Lourenço had lived in Goa, the center of diamond dealing in Portuguese India, he was accused by some of holding up confirmation of the discovery until he had lined his own pockets.

The initial code *(regimento)* regarding the diamond trade was drawn up by Dom Lourenço in 1730.[20] A head tax of 5 milreis was to be paid on each miner or slave seeking diamonds, and the *ouvidor,* or judge, at Villa do Principe was to decide on all questions concerning the distribution of *datas,* or claims. The code stressed the need for very strict supervision. No one was to be allowed to purchase diamonds from slaves; any friar found in the district was to be expelled; and there were to be no taverns or shops within two leagues of any mining activities. However, these restrictions did not satisfy the Crown when a few big diamonds which turned up in the Lisbon market were traced to Tijuco. From then on, the Brazilian authorities were ordered to limit all diamond washing to the Jequitinhonha and Ribeirão do Inferno waters. In addition, no claims were to be assigned: all were to be purchased at a rate of 60 milreis per square fathom (the length between two outstretched arms) or by the highest bidder over that sum. Furthermore, anyone not active in washing for diamonds was forbidden to live in the valleys even though they and their families might have been there prior to the discoveries. There had been gold camps in the region for nearly three decades, the chief center being Villa do Principe, and naturally a large number of families moved from there to Tijuco to establish themselves. Inevitably the wealth of the new town also caused small villages to spring up in the vicinity, and soon there were clusters of people at Rio Manso, Penha, Arassuahy, Rio Prêto, Govêa, Curimatahy, Pousa Alto, and elsewhere.

Dom Lourenço considered his orders somewhat extreme and therefore accepted a head tax of 15 milreis for each inhabitant in lieu of expulsion, and he allowed people to seek diamonds throughout the district if they would pay another 20 milreis per person. The edict prohibiting free Negroes and mulattoes from searching for diamonds was retained, however; in short, only whites and slaves were to work the region.

Upon the retirement of Dom Lourenço in 1732 the Crown decided that it was essential to appoint an Intendant in the Diamond District, and Dr. Raphael Pires Pardinho, a Crown attorney who had been active in the Captaincy of São Paulo, was selected for this post. Since he had judicial and fiscal powers as well as administrative responsibilities, he was by no means completely subordinate to the Governor of Minas Gerais or the Viceroy or Captain-General in Bahia or Rio de Janeiro. One of his first assignments was to establish the boundaries of the Sêrro do Frio region, and in this he was aided by a close friend of the King, Dom Martinho de Mendonça de Pina e Proença, who had served with Prince Eugene against the Turks and had also lived for a while in England. Together they drew a 50-league (approximately 150 miles) perimeter around Tijuco in the form of an elipse, the major diameter from north to south being 12 leagues and the shorter one from east to west, 7 leagues, and the district was separated sharply from the other settlements of the Captaincy. The probity of these two men did a great deal to change the undisciplined mining camps into law-abiding settlements. Their task was not an easy one, particularly as the European market was glutted shortly after their arrival and, as a result in 1734, diamond mining had to be prohibited for several years. Meanwhile the Count of Galvêas had been told by Lisbon to reduce the number of people in the region and to increase the head tax, which he raised to 40 milreis. In 1736, washings were renewed under tight supervision, and although the Intendant recommended that the Crown take over the project, make it a royal monopoly, and work it, Lisbon decided to farm out the task to contractors. Their actions, of course, were closely watched; no more than 600 slaves could be used in working the sites and there was a head tax payable to the Crown of 230 milreis on each Negro.

For three decades the Fernandes de Oliveira, Ferreira da Silva, and Caldeira Brant families operated the diamond concession, and during that period the Crown received 4,500,000 milreis from them. In 1751 it was decided that since gold was found with the diamond washings (the two always turned up together), a smelting house (casa da fundição) should be established in the region. Initially at Tijuco, the office was later transferred to Villa do Principe where the *ouvidor geral* was stationed. Under the Caldeira Brants the community prospered, handsome houses were erected, and costly churches built. There were festivals, elaborate music, and an intellectual life comparable to that of Villa Rica. Following the Lisbon earthquake of 1755 a good many Portuguese settled in the district, partly because there was such a demand for artisans, notably saddlers and blacksmiths, as well as doctors, apothecaries, barber-surgeons, school teachers, innkeepers, soldiers, and the like. Household goods, furniture, looking glasses, silver, Ming porcelain, clothing, calicoes, handkerchiefs, hats, and of course, tea, were imported in exchange not only for gold bars and precious stones but also for hides, cotton, marmalade, and cheese.[21]

In 1771 the Marquis of Pombal abolished the contract system and took over all mining for diamonds in the name of the Crown. The main feature of the new code was the establishment of a directorate of three members in Lisbon under the inspection of the General Director of the Treasury *(Real Erario),* who in turn named three administrative tellers to work in the Sêrro do Frio. Since the instructions arrived in a volume bound in green, it was known as the *Livro da capa verde* and always spoken of disapprovingly by the Mineiros. Some of the later eighteenth-century Intendants were extremely strict, and the role of the dragoons in controlling contraband was harsh. The Intendant Dr. José Antônio de Meireles, for instance, had the nickname "Cabeça Ferro" (Iron Head). Meanwhile, the clergy defended the interests of the inhabitants against the Crown authorities, and Padre Rolim, son of a mining entrepreneur, was the most prominent person from Tijuco in the Independence Movement of 1789, which is discussed below. In the 1790s the Intendant José Inacio do Amaral Silveira, known as "Coração de Ferro" (Heart of Iron), would not at first allow Dr. José Joaquim Vieira Couto to visit the Diamond District in order to make a scientific report for the Crown on precious stones and minerals found there.

Despite the stringent policy in the Sêrro do Frio, the region prospered and the remittances to Portugal did not decline in the last quarter of the century as they did in the gold-producing parts of the Captaincy. Traveling in the first decade of the nineteenth century, the English mining expert John Mawe was impressed by the standard of living in Tijuco:

> The residences of the upper classes are more attractively laid out and furnished than those which I saw in Rio de Janeiro and São Paulo. . . . In no part of Brazil did I meet with a society so select and agreeable. This may certainly be called the court of the mining district. In their manners there was no ceremonious reserve or courtly refinement but their whole demeanor was genteel and well bred.[22]

In retrospect, it is well to remember that for practically 200 years after Cabral's landfall, Brazil was considered primarily an agricultural colony. In the second decade of the nineteenth century, however, the German scientists Spix and Martius could think of it also as a region of great mineral deposits with excellent prospects for development:

> Almost every kind of metal is found; iron-stone which produces ninety per cent [iron], is met with nearly everywhere, and it constitutes in a manner the chief component part of long chains of mountains; lead is found beyond the Rio de São Francisco in Abaité; copper in São Domingos, near Fanado in Minas Novas; chrome and manganese in Paraöpeba; platinum near Gaspar Soares, and in other rivers; quicksilver, arsenic, bismuth, antimony and red-lead ore, about Villa Rica; diamonds in Tijuco and Abaité; yellow, blue, and white topazes, grass and bluish green aquamarines, red and green tourmalines, chrysoberyls, garnets and amethysts, principally in Minas Novas. But what has chiefly contributed to the great influx of settlers, and to the rapid population of this Captaincy [Minas Gerais], particularly of the capital [Villa Rica] is the great abundance of gold which has been obtained for above a century.[23]

The gold and diamond rush had many different effects on Brazil. Basically the economy was brought to life as precious metals provided the medium of exchange for the colony to power the wheels of its internal growth, finance imports, and pay taxes owed the Metropolis without being bankrupted or stripped of its circulating medium. Once the initial investment of production had been retrieved, gold and diamonds furnished the capital to support and rejuvenate agricultural and commercial activity on the coast. It attracted a sizable wave of immigration from Europe and brought in more slaves from Africa. These population movements affected both the economic life and the racial mix of Brazil. One of the most important results of the finds of precious metal and gems in the central *altiplano* was the way that it pulled the country together and acted as a hub for the spokes of the wheel. The littoral was now definitely linked with the *sertão;* transportation was possible by land or river not only from Pernambuco, Bahia, Rio de Janeiro, and São Paulo to Minas Gerais, Goiás, and Mato Grosso, but also from Belem do Pará at the mouth of the Amazon to the "Continent of São Pedro" (today Rio Grande do Sul). The internal market opened up trade for all parts of Brazil, and the eyes of the colonists could look elsewhere than to Africa and Europe. Instead of being solely a country with a narrow strip of land along a very long coast, there was now a hinterland. While the littoral remained of paramount importance, a drive toward western Brazil was now feasible with the opening of the vast sertão whose riches had been laid bare by the bandeirantes. Minas Gerais became a wealthy and populous state and today is one of the most important in the country.

For a number of reasons, diversified mining did not emerge on a large scale during the Imperial period, and Brazil's local and national economy had to function and grow in other ways. Agriculture took over, aided primarily by coffee, and mining waited until the twentieth century for a new surge. Modern Brazil's bold and gigantic project of building the new city of Brasília is an effort to weld the nation into a still more integrated entity. The idea of a capital in the interior was advocated by the great scientist and statesman José Bonifacio in 1821, and it was put into effect in 1957 by President Juscelino Kubitschek, who was born in Diamantina and was a four-term Governor of Minas Gerais. Today Brazil is one of the world centers for industrial metals. Its high-grade iron-ore reserves are estimated at more than 13 billion tons and its manganese deposits are the largest in the Western Hemisphere. It is a major source for industrial diamonds and the chief source for quartz crystals for use in radios. If Brasília does succeed in giving greater cohesion to a land whose scattered populations have sometimes seemed like islands in a large sea, part of the reason will be due to the healthy continuation of its mining industry.

Notes

1. The question whether this was due to accident or design has been much debated. Edgar Prestage, *The Portuguese Pioneers* (London: Black, 1933), and Fidelino de Figueiredo, "The Geographical Discoveries and Conquests of the Portuguese," *Hispanic American Historical Review,* vol. VI (1926), 47–70, together with C. E. Nowell, "The Discovery of Brazil—Accidental or Intentional?" ibid., vol. XVI (1936), 311–338, argue for intention. Samuel Eliot Morison in *Portuguese Voyages to America in the Fifteenth Century* (Cambridge, Mass.: Harvard University Press, 1940) and W. B. Greenlee in *The Voyage of Pedro Alvares Cabral to Brazil and India, from Contemporary Narratives and Documents,* Hakluyt Society Publications, ser. II, no. 81 (London, 1938), believe the discovery was accidental.

2. Charles R. Boxer, *The Dutch in Brazil, 1624–1654* (Oxford: Clarendon Press, 1957); also his *Salvador de Sá and the Struggle for Brazil and Angola, 1602–1686* (London: University of London, Athlone Press, 1952).

3. William Dampier, *Dampier's Voyages,* John Masefield, ed., 2 vols. (London, 1906), vol. II, p. 381.

4. See the collection of readings and bibliography concerning this social group in Richard Morse, *The Bandeirantes* (New York: A. A. Knopf, 1965). Morse says the earliest document using the word *bandeirante* dates from 1740. Ibid., p. 23.

5. Fernão Cardim, *Tratados da terra e gente do Brasil: Introduções e notas de Baptista Caetano, Capistrano de Abreu e Rodolfo García,* 2d ed. (Rio de Janeiro: Brasiliana Bibliotheca Pedagógica Brasileira ser. 5a, 1939), vol. 168, p. 295. Also cited in Samuel Putnam, *Marvelous Journey: A Survey of Four Centuries of Brazilian Writing* (New York: A. A. Knopf, 1948), p. 48.

6. Antônio Vieira, "Sermão da primeira oitava da Páscoa," in *Obras Escolhidas,* Antônio Sergio and Hernani Cidade, eds. (Lisbon, 1954), vol. XI, p. 275.

7. Pedro Calmon, *História do Brasil: Século XVI,* 2d ed. (Rio de Janeiro, 1963), vol. I, pp. 181 and 183.

8. Toribio de Ortiguera, *Jornada del Río Marañon* in Nueva biblioteca de autores españoles, no. 15, M. Serrano y Sanz, *Historiadores de Indias* (Madrid: Bailly–Bailliére é Hijos, Editores, 1909), vol. II, pp. 305–422.

9. Gabriel Soares de Sousa's *Tratado descritivo do Brasil em 1587* has appeared in a number of modern editions, the best text being based on the 1851 edition of Varnhagen. See José Honorio Rodrigues, *Historiografia del Brasil: Siglo XVI* (Mexico City, 1957), pp. 47–56. Soares went to Madrid to obtain donatarial privileges, since Brazilian affairs were handled by a special Portuguese Council which attended the Court of Philip II (I of Portugal) in Spain.

10. The French explorer Charles Marie de La Condamine mentions having seen this document, dated August 23, 1639, in the archives at Pará in the middle of the eighteenth century. In it Teixeira declares that he placed a landmark at the spot which he called Villa del Oro—actually the mouth of the Napo—where he exchanged trinkets with the natives for gold objects as a sign of having taken possession of this region for the Portuguese Crown. La Condamine adds that this news was confirmed a few years afterward in the diary kept by Fathers Acuña and Artieda who accompanied the expedition on its return from Quito to prepare a report for Philip IV (III of Portugal). In it they declared that they had witnessed the inhabitants of the Yupur and Iquiari Rivers trade in this metal with their neighbors. For the diary see *Viaje a la América meridional,* Colección Austral (Madrid, 1944).

11. Manoel Cardozo, "Dom Rodrigo de Castel-Blanco and the Brazilian El Dorado, 1673–1682," *The Americas* (Franciscan Academy), vol. I, no. 2 (October 1944), pp. 131–159.

12. Much of the material for this chapter has been drawn from João Antônio Andreoni (pseud. Antonil), *Cultura e opulência do Brasil por sus drogas e minas* (Lisbon: Na Officina Real Deslandesiana, 1711) reprinted with a French translation and excellent notes by Andrée Mansuy (Paris: Institut des Hautes Études de l'Amérique Latine, Université de Paris, 1968); Charles R. Boxer, *The Golden Age of Brazil, 1695–1750* (Berkeley: University of California Press, 1962); Manoel Cardozo, "The Brazilian Gold Rush," *The Americas* (Franciscan Academy), vol. III, no. 2 (October 1946), pp. 137–160; Bailey Diffie, *Latin American Civilization: Colonial Period* (Harrisburg, Pa.: Stackpole Sons, 1945), book III; and Celso Furtado, *The Economic Growth of Brazil: A Survey from Colonial to Modern Times,* R. W. de Aguiar and E. C. Drysdale, trans. (Berkeley: University of California Press, 1965).

13. For this and the next citation, see Antonil, *Cultura e opulência do Brasil,* part III, "Pelas Minas de Ouro," chap. II, par. 2, pp. 350–355 of the Mansuy edition.

14. Antonil, op. cit. chap. V, pp. 366–367, and Boxer, *Golden Age,* p. 49.

15. Antonil, op. cit. chap. VI, pp. 376–379.

16. The thesis of Portugal's early indecision regarding the gold strikes of Minas Gerais is drawn from Cardozo, op. cit.

17. Antonil, op. cit. chap. V, pp. 368–369.

18. The title of the book in the Portuguese original is *Triunfo Eucharistico exemplar de Christandade Lusitana . . . em Villa Rica, corte de capitania das Minas . . . Dedicado a soberana senhora do Rosario pelos Irmãoes Prêtos de sua irmandade, e a instancia dos mesmos exposto a publica noticia por Simam Ferreira Machado natural de Lisboa, e morador nas Minas* (Lisbon, 1734). There is a facsimile reprint in Alfonso Ávila, *Residuos seiscentistas em Minas: Textos do século de ouro e as projecções de mundo barroco* (Belo Horizonte, 1967), p. 24. See also Boxer, *Golden Age,* pp. 162–163. Another interesting contemporary work is Luis Gomes Ferreira's *Erario mineral dividido en doze tratados . . . Autor L.G.F. cirurgião approvado, natural da Villa de S. Pedro de Rates, e assistente nas Minas do ouro por discurso de vinte annos* (Lisbon, 1735), which is important for the social history of the period. One of the best overall modern studies of the region in colonial days is João Camillo de Oliveira Torres, *História de Minas Gerais,* 5 vols. (Belo Horizonte, 1961).

19. The principal studies on the Diamond District are Felicio dos Santos' mid-nineteenth-century classic, *Memórias do Distrito Diamantino da Comarca de Sêrro Frio,* 3d ed. (Rio de Janeiro, 1958); Augusto de Lima, Jr., *A Capitania de Minas Gerais* (Lisbon, 1940); Aires da Mata Machado Filho, *Arraial do Tijuco, cidade Diamantina* (Rio de Janeiro, 1945); Augusto de Lima, Jr., *História dos diamantes nas Minas Gerais* (Lisbon and Rio de Janeiro, 1945). See also Boxer, *The Golden Age,* especially chap. VIII.

20. For the Regimento dos Diamantes (June 1730) see A. de Lima, Jr., op. cit., pp. 32–39.

21. Queen Catherine of Braganza, the wife of Charles II, was responsible for introducing tea to England after the Restoration, and the poet Waller was truly grateful:

    The best of Queens and best of herbs we owe
    To that bold nation who the way did show
    To the fair region where the sun doth rise,
    Whose rich productions we so justly prize.

    The oriental quince, long familiar in the Mediterranean, was planted in the hinterland of Brazil, and its Portuguese name, *marmelo,* resulted in the international word marmalade. Brazilians also adapted the process to the fruit of the guava *(goiaba)* tree, which is native to South America, and *goiabada* or guava paste was equally popular. Of parallel importance was the making of cheese from goats' and cows' milk in the mining region. Queijo-de-minas—a white, dry, slightly salted cheese which can be kept almost indefinitely—became famous in the eighteenth century, and various other types were also developed. The semicured type from the Sêrro do Frio and Diamantina was and is particularly appreciated.

22. John Mawe, *Travels in the Interior of Brazil* (London, 1812), pp. 230–231. There is also an edition in Portuguese, *Viagens ao interior do Brasil,* Solena Benevides Viana, trans., Clado Ribeiro Lessa, notes (Rio de Janeiro, 1944).

23. John B. von Spix and C. F. P. von Martius, *Travels in Brazil in the Years 1817–1820 Undertaken by Command of His Majesty The King of Bavaria,* H. E. Lloyd, trans., 2 vols. (London, 1824), vol. II, pp. 181–182. The Portuguese translation by Lúcia Furquim Lahmeyer is entitled *Viagem pelo Brasil,* B. F. Ramiz Galvão e Basílio de Magalhães, rev., Basílio de Magalhães, notes, 4 vols. (Rio de Janeiro, 1938).

VI

El Dorado: Tale and Truth

Humboldt states that according to official Treasury records from 1545 to 1803, the mines of the Cerro de Potosí alone produced silver valued at 1,095,500,000 pesos[1]:

Official Totals of Silver Produced at Potosí, 1545–1803 by Periods

1545 to 1556	127,500,000 pesos
1556 to 1789	819,258,500 pesos
1789 to 1803	46,000,000 pesos
Total	1,095,500,000 pesos

Florisel, following Humboldt's figures, arrives at a figure of 4,400,-000,000 pesos of eight reales ("pieces of eight") for the total value of gold and silver sent to Spain from all of America in the years between 1492 and 1803, including the sums stolen by pirates or lost at sea and including Humboldt's estimate of metals illegally mined and therefore not registered. He makes the following calculations:

> If we divide 4,440 million pesos into the 311 years of Spain's guiding and civilizing hand in America, they yield a quotient of a little more than twelve million per year. . . . America gave twelve million pesos annually in exchange for all the good and all the possible knowledge of the Cross and the wisdom of the world and of History. This knowledge of wisdom, collected, cultivated, and nourished at the time in Castilian hearts, was generously transfused, in all its living integrity, into the virgin mold of the New World . . . like a gift of the gods.[2]

Florisel's reasoning is a defense against the common but fallacious argument that Spain criminally exploited America without giving her anything in return. It is not possible to pose a proposition in such terms, as if it were a sort of commercial transaction, a give-and-take business deal between two parties in which one profits at the expense of the other. The discovery of the New World, and the consequent creation of the Spanish- and Portuguese-speaking nations who occupy it today, is an event of historical destiny and must be regarded as such. Spain and Portugal settled the conquered lands according to the temper of their people and within the framework of existing economic and political ideas and circumstances. And it fell to mining to be the primary instrument of that transcendental happening. The final results cannot be presented on a balance sheet in double-entry bookkeeping.

Under the dominant economic philosophy of the age—mercantilism—a nation's wealth, and hence its strength, was to be measured primarily by its stocks of gold and silver, for with those metals anything could be bought: Flemish laces and warships, fine textiles and cannon, foodstuffs and excellent mercenaries. As Columbus put it so simply in his letter from Jamaica in 1503: "Gold is a marvel! He who possesses it is owner of all that he desires. With it even souls can be raised into Paradise." Unlucky nations not possessing mines at home or in their colonies had to manufacture goods for export or trade for precious metals—or even steal from their more affluent neighbors. But Spain and later Portugal were luckier than other nations; the mines of New Spain, Peru, New Granada, and Minas Gerais gave them access to untold treasure.

Gold and silver gave the Iberian countries political power and importance in the turbulent world of the time, but only precariously and transitorily. Perhaps the power and importance were fleeting precisely because of the influx of the precious metals which the Conquest supplied in unimaginable quantities. Since Spain's merchants and manufacturers could not meet new national responsibilities, and the Casa de Austria desperately needed specie to pay for its wars and maintain the Holy Roman Empire, American gold and silver tended to pass quickly through the Peninsula with little advantage to the nation. In a sense, this silver and gold laid the base for the disintegration of Spain's economy due to the corrosive influence of the excess of money in circulation. The ultimate destination of the metals mined in Guanajuato and Potosí, in Zacatecas and on the Sinú River, and sent to Spain in convoys sailing from Veracruz, Cartegena de Indias, or Portobello, was not Spain itself. Instead, the gold and silver made its way finally to cities such as Antwerp, Bruges, Ghent, Amsterdam, London, Hamburg, Bremen,

Genoa, and Basel—the money markets where commercial capitalism was spawned and maintained.

By a curious irony, those same money markets which received the metals so carefully registered in the House of Trade in Seville also received the wealth obtained by pirates through assault and rapine. Without a doubt American precious metals had a decisive influence on the development of modern capitalism and the culture of Western Europe.

Not all of the gold and silver extracted in the New World crossed the Atlantic, however. It also circulated in large amounts within the American continent itself, where it was very important in the development of the economy. In addition, it is a little-known fact that, together with the Royal Exchequer, the Viceroyalties of New Spain and Peru were responsible for contributing from their own treasuries to cover the deficits occasioned by the support of broad areas beyond their own jurisdictions. Shipments of metal or minted coins for this purpose were called *situados.* Peru sent situados to Chile and the Río de la Plata region, while through New Spain the Captaincies-General of Yucatán and Guatemala, gave financial aid to all of Central America and as far away as Panama, as well as to Florida, Cuba, Puerto Rico, and Jamaica (as long as that island remained in Spanish hands). Furthermore, throughout the period of the Viceroyalties Mexico was responsible for the political and economic needs of the Philippine Islands.[3]

No one will ever know the amounts of gold and diamonds mined in Brazil. Contraband was widespread and royal authority diffuse, so that the Crown's Fifth was obviously not what it should have been. Substitute measures such as head taxes on slaves and fixed sums from each miner also were difficult to enforce. The distinguished economic historian of the colonial period, João Pandiá Calógeras, estimated the total gold produced from 1700 to 1801 at 65,500 arrôbas, or 983,000 kilograms (some 2,162,600 pounds avoirdupois). He set the output of Mato Grosso and Goiás between 1720 and 1801 at 13,000 arrôbas and the outlying areas of São Paulo and Bahia-Ceará at 5,000 arrôbas.[4] The years of greatest productivity probably were those between 1740 and 1760.[5]

Portugal's wealth from minerals followed a pattern similar to that of Spain although, because of the Methuen Treaty signed with England in 1703, the bulk of Portuguese trade was with her old

Since they had no machine that could sort diamonds from pebbles, the Brazilians used slaves to perform this task. The illustration is from John Mawe, *Travels in the Interior of Brazil* (1812). (*The New York Public Library.*)

The remarkable sanctuary of Bom Jesus de Matozinhos at Congonhas do Campo in Minas Gerais was erected through the efforts of a miner turned hermit who gathered funds from both rich and poor residents and visitors in the wealthy mining district. (*Organization of American States.*)

The Church of Bom Jesus is most famous for its series of statues of the Twelve Prophets created by the crippled native Brazilian sculptor Aleijadinho, one of the most extraordinary artists of the New World. (*Organization of American States.*)

A stylized rendering in Théodore de Bry's *Americae* (1602) depicts llamas carrying silver from the mines of Potosí to the coast along an ancient route of the Incas. Such caravans were scarcely guarded during the two-week-long trip since there was little danger of robbery. (*Rare Book Division, The New York Public Library, Astor, Lenox and Tilden Foundations.*)

Total and Average Annual Production of Gold in Minas Gerais, 1700–1801

Period	Number of Years	Total Production, arrôbas	Average Annual Production, arrôbas
1700–1725	26	7,500	288
1726–1735	10	6,500	650
1736–1751	16	12,000	750
1752–1787	36	18,000	500
1788–1801	14	3,500	250

ally. While her wines enjoyed special privileges over those of France, other manufactures declined in Portugal, and Brazilian gold went to London to purchase foreign goods. As the historian Southey remarked: "Brazil was supplied almost exclusively with English manufactures through Portuguese merchants of the capital, to whom the members of the British Factory gave two or three years credit."[6]

The international effect of the remittances from Minas Gerais was profound. As Afonso Taunay pointed out:

> The infusion of great streams of Brazilian gold into European markets exerted a very strong influence on the economy of the Western World. During the eighteenth century this had already been noticed by authorities as important as Montesquieu in his *De l'Esprit des Lois* (1748) and Adam Smith in his famous *Inquiry into the Nature and Causes of the Wealth of Nations* (1776). In 1728 the first of these great authors had called public attention to the abundance of Brazilian gold circulating in civilized Europe as exercising beneficial influence on the economy of the northern part of the continent. Adam Smith emphasized how much the plentiful Brazilian metal had increased Anglo-Portuguese commercial exchange. He went so far as to admit that in his time almost all the gold coined in England came from Brazilian deposits. . . . Another famous and, so to speak, modern authority, Werner Sombart, considered that without the discovery of the gold deposits of Brazil the development of modern economic man would have been delayed. It was the Brazilian metal which enabled Great Britain to build up the large reserves which assured her predominance in world commerce, as long-lasting as it was remarkable, through the progress and improvement of her industries.[7]

Fortunately a great deal of silver and gold remained in the New World, and generally speaking, the colonists prospered more than the majority of the people in the mother countries. That was one of the reasons, of course, for the continuous emigration. The larger cities could hold their own with those of the Peninsula, and there were thousands of churches and monasteries which were handsome, spacious, and especially rich in decoration. Indeed, in many respects they outdid those in the Old World. Certainly a far greater number of churches were erected in Spanish and Portuguese America than that in the Peninsula during the same period.

The relationship between Brazil and the Spanish-speaking Viceroyalties was a factor which should not be ignored in this discussion. Even before gold was found in Minas Gerais, the silver of Alto Peru stimulated Portuguese America during the "Babylonian Captivity," and trade in sugar, slaves, and goods from India and other distant lands flourished between Pernambuco and Bahia and Buenos Aires. La Plata, Potosí, and Lima were thus supplied "through the back door," and coins of four and eight reales were in circulation throughout Brazil. Much of the trade was clandestine since, until the Viceroyalty of Buenos Aires was set up in 1776, the Spanish Crown tried to have all commerce concentrated on the west coast. The settlers, however, were more than anxious to deal with one another, and Cassiano Ricardo has pointed to the number of Spaniards who were active in São Paulo and who took part in Brazil's "gold rush."[8] Although the supply of gold in Portuguese America diminished after 1760, diamonds remained important and the momentum of the discoveries continued. As Diffie sums it up:

> Gold and diamonds rescued Brazil from economic decay and stimulated her agriculture, the rise in sugar production continuing upward with the exports of precious metals and stones. . . . The latter eighteenth and early nineteenth centuries were far from being a period of decline. . . . The agricultural exports of Brazil tripled in the last quarter of the eighteenth century, Rio exporting about one-third of the total and the ports from Bahia northward about two thirds. . . . The last half century of Brazil's colonial history saw the climax of the economic competition between north and south Brazil. North Brazil, mainly Bahia and Pernambuco, was the colonial leader. Gold and diamonds brought Rio forward, and at the end of the colonial period the three ports were about on a level, one exporting about the same as the other with almost yearly changes of leadership.[9]

Notes

1. Alexander von Humboldt, *Essai politique sur la Royaume de la Nouvelle-Espagne,* book III, chap. 11, but with figures summarized by Florisel, *El cuento y la cuenta del oro de América* (Mexico, 1927), pp. 40 ff.

2. Ibid.

3. Arturo Arnaiz y Freg, "La evolución histórica," *Revista de Filosofia y Letras* (Mexico), 1948. Also see the interesting exchange of correspondence between Edward G. Bourne and James A. Le Roy in the *American Historical Review* for 1905 and 1906 entitled "The Philippine 'Situado' from the Treasury of New Spain," vol. X, pp. 459–461 and 929–932; vol. XI, pp. 722–723.

4. Bailey Diffie, *Latin American Civilization: Colonial Period* (Harrisburg, Pa.: Stackpole Sons, 1945), book III, pp. 679–681. João Pandiá Calógeras, *Formação histórica do Brasil,* 3d ed. (São Paulo, 1938), p. 54, and R. C. Simonsen, *História económica do Brasil,* 2 vols. (São Paulo, 1937), vol. II, pp. 93–96. A Portuguese *arrôba* was the equivalent of about 32 pounds. [Manoel Cardozo, "The Brazilian Gold Rush," *The Americas* (Franciscan Academy, vol. III, no. 2, (October 1946), p. 138 n. 11.]

5. Tabulation after Diffie, op. cit. p. 680

6. Robert Southey, *History of Brazil,* 3 vols. (London, 1810–1819), vol. III, p. 550.

7. Afonso d'Escragnolle Taunay, *História geral das bandeiras Paulistas,* 2 vols. (São Paulo, 1954), vol. II, 317–320, in Morse, trans., *The Bandeirantes* (New York; A. A. Knopf, 1965), pp. 187–188.

8. Lewis Hanke, "The Portuguese in Spanish America, with Special Reference to the Villa Imperial de Potosí," *Revista de Historia de América,* no. 51 (June 1961), pp. 4–5, and Cassiano Ricardo, *La marcha hacia el oeste* (Mexico, 1956), particularly chap. XIII: "El elemento español en la interpretación psicosocial de la bandeira," pp. 405–420. The second Portuguese edition, *Marcha para oeste,* 2 vols. (Rio de Janeiro), came out in 1942. See also Morse, op. cit., pp. 205–206, for several paragraphs translated from Ricardo on this subject. Vicente Licínio Cardozo's *A margem da história do Brasil* (São Paulo, 1933), pp. 79–101, stresses the Spanish influence in the growth of Brazil.

   There is a revealing letter of Manuel Juan de Morales of 1636 in Jaime Cortesão, ed., *Jesuitas e bandeirantes no Guaira (1549–1640), Manuscritas da Coleção de Angelis* (Rio de Janeiro, 1951), pp. 182–193; another pertinent study is Nilo Garcia, *Aclamação de Amador Bueno: Influência espanhola em São Paulo* (Rio de Janeiro, 1956). Finally, it should not be forgotten that Salvador de Sá, Governor of Rio de Janeiro in the mid-seventeenth century, was a brother-in-law of Luís de Cespedes, Governor of Paraguay. [Charles R. Boxer, *Salva-*

dor de Sá and the Struggle for Brazil and Angola, 1602–1686 (London: University of London, Athlone Press, 1952), pp. 82–100.]

9. Diffie, op. cit., pp. 689–690.

VII

Mining as a Stimulus to the Economy

In order to produce the enormous quantities of precious metals for which they were famous, the mining centers which dotted Ibero-America required great supplies of clothing and utensils, tools and machinery, pack animals and carts for mine operations and transport, and homes for the miners. They also needed lines of communication and routes to the outside world. This demand meant that lands were cultivated from the very beginning of European occupation and that trade and industry were established and rudimentary roads built. In addition, specialized workshops were set up for those trades whose development followed the fortunes of mining.

It often has been said that mining operations hindered the development of agriculture everywhere in America. The truth is quite the opposite, except in those regions totally unsuitable for cultivation, such as the land in the cold region of Upper Peru. Humboldt, as always, made shrewd observations on this subject:

> It cannot be observed that agriculture is more neglected in Peru than in the province of Cumaná or in Guiana, especially since these last have no mines in operation. In Mexico the best-cultivated fields, those which remind travelers of the most beautiful countrysides of France, are the plains extending from Salamanca to the neighborhood of Silao, Guanajuato, and the Villa de León, and which contain the richest mines in the known world. In all the places where veins of ore have been discovered, in the wildest parts of the mountain ranges, in remote and desert-like tablelands, the exploitation of mines, far from impeding cultivation of the soil, has signally stimulated it! Travels over the backbone of the Andes or in the mountainous parts of Mexico offer the most striking examples

> of this beneficent influence of the mines on agriculture. . . . Without the establishments built for the exploitation of the mines, how many places would have remained deserted, how many lands would not have been cleared for cultivation in the four intendancies of Guanajuato, Zacatecas, San Luis Potosí, and Durango, between the twenty-first and twenty-fifth parallels, where the most considerable part of New Spain's mineral wealth is concentrated! The founding of a city follows immediately on the discovery of an important mine. If the settlement is located on the arid flank or crest of the ranges, the new settlers will have to travel a long distance to seek everything they need for their own subsistence and that of a large number of animals employed in pumping out water, in the extraction and amalgamation of the ore. Necessity at once brings industry into being: men begin to work the soil in the ravines and slopes of the neighboring mountains in every place where the rock is covered with humus. Ranches are established in the vicinity of the mines. The dearness of foodstuffs and the high price level at which all the products of agriculture are maintained, owing to the large numbers of consumers, compensate the farmer for the privations to which the toilsome life of the mountains exposes him. . . . Indeed, this influence of mines in the progressive agriculturalization of the countryside is more lasting than the mines themselves. When the lodes are exhausted and the subterranean workings abandoned, there is no doubt that the population of the district declines, for the miners go to seek their fortunes elsewhere; but the farmer is bound by the attachment he has formed to the soil that has witnessed his birth, and which his parents cleared with their own hands.[1]

In fact, many rich miners spent part of their fortunes on agricultural exploitation. In regard to New Spain, Don Lucas Alamán tells us:

> From the wealth of the mines of Zacatecas came the opulent homes of the Counts of San Mateo Valparaíso, Santa Rosa, Santiago de la Laguna, and many other newly enriched men of Bolaños (the Marquises of Vivanco), of Sombrerete (the Fagoaga family) and, in the present-day State of Hidalgo, Real del Monte, of the Counts of Regla; these mines were the starting point of many of the wealthy families of the region, who became the chief landowners.[2]

And, adds Alamán:

> The large sums of money poured out by mining establishments spread for many leagues around, developed agriculture and industry through the demands made on the products of

> both, which were produced for the working, draining, and extraction of metals: the religious spirit of that century, and even the prodigality for which the miners were noted, contributed to this development, for mine owners spent a part of their profits, never counting the cost, on construction of churches, monasteries, and hospitals; and the miners themselves, who at that time worked on shares in all the mines, were lavish in spending the money they received for their labors. This caused the value of rural properties to rise sharply within the area most directly influenced by the mines, and thus we see that in the Guanajuato region the value of ranches, and the incomes they produced, doubled and more than doubled in the course of a few years.[3]

The same could be said of livestock breeding, which provided an indispensable element in the inhabitants' food supply as well as animals for work in the mines and mine yards and for the transportation of ore and refined metals. Growth in numbers of all kinds of livestock imported from Spain was both early and rapid. The varieties ranged from the horse, which had arrived on the continent with the first conquistadors, to the mule and donkey, and to the cattle which had arrived with Columbus. They also included all types of smaller animals such as sheep, goats, and swine. In the words of the historian Faustino Miranda:

> Livestock covered Mexican soil like a flood. By the end of the sixteenth century it was frequent to encounter herds numbering ten or twenty thousand head. In the center of the country the most suitable areas for maintaining herds overflowed the region, and the restrictions placed on them were barely sufficient to prevent their spilling over into agricultural lands.[4]

The close, but often unappreciated, relation between mining and agriculture, and the resultant benefits to both, was clearly pointed out by the historian Don Carlos Pereyra:

> Taking the question as a whole, no American country, not even Mexico, was more of a mining than an agricultural country. The greater the mining activity, the more progress was made in the area of agricultural production, as was demonstrated in Guanajuato, the Lombardy of America as well as being the most important mining center in the world at the end of the eighteenth century. There, the highest wages were paid, and the soil was cultivated by large agricultural entrepreneurs.[5]

The explanation for the development of trade within the New World is obvious: it was necessary to satisfy the needs of the scattered settlements occupying a large undeveloped region. As a consequence, long roads, whose construction appeared incredible by the standards of the time, had to be built to accommodate the ever-increasing traffic of pack animals. The creation of ports and the building of ships also were part of the new activity. All of this culminated in the transportation to the Iberian Peninsula of precious metals from the remote regions where they were mined.

Don Antonio de Mendoza, the first Viceroy of New Spain, built a road passable for pack animals between Tasco and Sultepec. In 1542, Fray Sebastian de Aparicio began the cart road from Mexico City to the mining town of Zacatecas, which was not to be finished until 1570. Eventually that road led through the central highlands of Mexico as far north as Durango. José de la Borda, while working some mines in Zacatecas, improved the Acapulco road in 1750 by routing it through Tasco. Humboldt speaks of the advantages of "the superb road which the *Consulado* [Merchants' Guild] of Veracruz is having built from that city to Perote. It can compete with those of the Simplon and Montcenis."[6] Humboldt says of this road, which was finished in 1803, that its gradient was very gentle and that it would accommodate travel by carriage, aid commerce, and considerably lower the cost of merchandise. The rise in highway traffic naturally coincided with the increase in mining wealth. Suffice it to say, in regard to New Spain, that in 1811, at the height of the Independence Movement, the road from Veracruz to Mexico City alone was used by 67,871 mules, 145 carriages, and 1,000 litters.[7]

In Peru, the excellent highways built by the Incas did not reach as far as the mines, which, generally speaking, were not discovered until the post-Columbian colonial period. The Spaniards neglected many of the Inca-built roads because they "ran the wrong way" for the new technology. Since the sea was not an important part of the Incan culture, the Incas built roads running north and south to knit together their 3,000-mile long empire; few roads ran west to east, from the coast into the sierra. In any event, the Incan roads were built primarily as paths or footwalks for the movement of troops, porters, and llama-trains since they had no wheeled vehicles.

The Spaniards, on the other hand, were sea-oriented. Their lines of communication and transport ran from the mine, the source of the product, to the Pacific and then by ship to Lima or, if bound for Europe, to Panama. José de Veitia described such a shipment in the seventeenth century:

> A report of September, 1606, states that a shipment of silver carried on the backs of mules or llamas spent two weeks on the road from Potosí to Arica, on the Pacific coast; from there to Callao [the port city for Lima] by ship required a week, and from that port to Panama a voyage of twenty days, [the ship] taking on silver from Paita and Trujillo en route.[8]

Panama was the great entrepôt for European goods destined for the west coast of South America and the Río de la Plata. Goods from these points were shipped north to be exchanged for European imports at the great fair at Portobello which was held to meet the fleet and its convoy of galleons. Father Thomas Gage has left us a description of that fair in 1637:

> [Panama] is held to be one of the richest places in all America, having by land and by the river Chagres commerce with the North Sea, and by the South, trading with all Peru, East Indies, Mexico, and Honduras. Thither is brought the chief treasure of Peru in two or three great ships, which lie at anchor at Puerto de Perico some three leagues from the city. . . .
>
> But what I most wondered at was to see [in Portobello] the *requas* of mules which came thither from Panama, laden with wedges of silver; in one day I told two hundred mules laden with nothing else, which were unladen in the public marketplace, so that there the heaps of silver wedges lay like heaps of stones in the street, without any fear or suspicion of being lost.
>
> It was worth seeing how merchants sold their commodities, not by ell or yard, but by the piece and weight, not paying in coined pieces of money, but in wedges which were weighed and taken for commodities. This lasted by fifteen days, whilst the galleons were lading with wedges of silver and nothing else; so that for those fifteen days, I dare boldly say and avouch, that in the world there is no greater fair than that of Portobello, between the Spanish merchants and those of Peru, Panama, and other parts thereabouts.[9]

Father Gage also says that later this traffic did not come by way of Panama and Portobello, but through Cartagena de Indias, from New Granada or Colombia. Portobello was so unhealthy that every effort was made to avoid long stays in its port by the fleets and those on board the ships, which stopped only in passing.

In time, a north-to-south land route 1,850 leagues long extended from Cartagena to Bogotá, Quito, Lima, Cuzco, La Paz, Potosí, Jujuy, and Córdoba and ended at the shores of the Río de la Plata at Buenos Aires. While the portion north of Lima was developed as an alternate line to the sea route for safer communication and to link various inland points, the Lima–Buenos Aires portion was developed to give the La Plata region a tie to Spain via Lima and Panama! This seemingly mad piece of planning had an inner logic of its own. The development of population centers on the Río de la Plata and around Asunción called for the establishment of a route to Seville. At the time, privateering and buccaneering made unprotected ships carrying silver an invitation to capture; commerce between Spain and her colonies had to be carried on under the guard of galleon convoys. Since convoys were expensive to mount, and the population and mineral wealth of the La Plata region were both scarce, legal trade from Spain directly to Buenos Aires and Asunción was restricted to one ship a year to trade European goods for local products, mainly hides. The road from Buenos Aires to Potosí was blocked by a customs house at Córdoba to keep Potosí's silver flowing toward Lima and to keep Buenos Aires's imports away from the interior. While Potosí silver was not allowed by law to go to Buenos Aires, it could be shipped south as far as the region west of Córdoba (particularly the Salta-Jujuy region of northwest modern Argentina) to pay for foodstuffs; for livestock for meat, leather, and draft animals; and for crude manufactures which could not be produced in the bleak mountains. Once this silver started southeast with the legitimate merchants' caravans, it was but a short step for it to continue illegally past Córdoba to serve as a medium of exchange for La Plata, purchase goods in Buenos Aires for Potosí, and eventually be shipped to Spain. In the eighteenth century, as piracy declined and Portuguese, British, and Dutch smugglers set up bases close to Buenos Aires to offer goods at "reasonable" prices, the Spanish Crown had to rethink its policy. At last, despite the strength of the merchants of Lima and Panama, the Crown reorganized the southeastern portion of the continent politically and recognized economic realities.

Setting up the Viceroyalty of La Plata in 1776 was a major change in Spanish colonial policy. This political unit not only comprised modern Argentina, but Uruguay, Paraguay, and most of Bolivia as well. Its economic raison d'être was to blend the expanding export of hides from the pampas with the dwindling export of silver from Potosí while collecting the revenue due the Crown and stamping out smuggling in the La Plata region. Although traffic along the entire 1,200 leagues of highway from Buenos Aires to Lima increased—since getting to Peru through this back door rather than over the Panamanian isthmus was often the more desirable route—the stretch from Potosí to Buenos Aires was by far the busiest. This road, cutting diagonally across the South American continent, has been delightfully described by Alonso Carrió de la Vandera in his book, *A Guide for Inexperienced Travelers Between Buenos Aires and Lima,* published in 1773 under the pseudonym "Concolorcorvo." Carrió's descriptions of the social life and customs along the road are matched by his interest in economic matters.[10]

Over the Lima to Buenos Aires road moved the merchandise of the Empire, and the market at Potosí was one of the great commercial centers on the route. The sale of merchandise was so profitable that Spaniards, otherwise too proud, became merchants, and some even became mere grocers. Because of the stigma attached to trade, some of these men passed off their commercial trips as hunting parties! Dealing with the free-spending miners made possible the amassing of a fortune in a short time. And so, the fame of the great fair at Potosí spread over the world and soon attracted large numbers of eager merchants.[11] Bartolomé Arzans de Orsúa y Vela, the chronicler and historian of Potosí, gives the following list of goods that flowed in from the far corners of the earth:

> silks of all kinds, and knitted goods from Granada; hose and swords from Toledo; clothes from other parts of Spain; iron from Viscaya; rich linen and knitted goods from Portugal; fabrics, embroideries of silk, gold, and silver, and beaver hats from France; tapestries, mirrors, elaborate carved desks, embroideries and laces from Flanders; clothes from Holland; swords and other steel implements from Germany; paper from Genoa; silks from Calabria; hose and fabrics from Naples; satins from Florence; clothes, embroideries, and fine fabrics from Tuscany; silver and gold lace and clothes from Milan; paintings and sacred pictures from Rome; hats and woolen cloth from England; glass from Venice; white wax from Cyprus, Crete, and the Mediterranean coast of Africa; scarlet cloth, glass, ivory, and precious stones from India; diamonds from Ceylon; perfumes from Arabia; rugs from Persia, Cairo,

> and Turkey; all kinds of spices from Malaya and Goa; porcelain and silk cloth from China; Negroes from the Cape Verde Islands and Angola; cochineal, vanilla, cocoa, and precious woods from New Spain and the West Indies; pearls from Panama; rich cloths from Quito, Riobamba, Cuzco, and other provinces of the Indies; and various kinds of raw materials from Tucumán, Cochabamba, and Santa Cruz.[12]

Districts hundreds of miles around vied with each other to provide supplies and Portuguese merchants from Brazil—called *peruleros,* from the word Peru—were thoroughly familiar with the route to Potosí![13] Chilean horses fetched fantastic prices, for they were considered "so spirited that in truth they could be compared to the zephyrs of far-famed Betis."[14] Raúl A. Molina, an Argentine economic historian, has described the Potosí–Buenos Aires route as the axis around which turned the whole political economy of the Río de la Plata or possibly the entire Empire.[15]

As far as New Spain was concerned, precious metals were the major merchandise assuring Mexico of a lively foreign trade with Europe and the Far East. According to Gustavo P. Serrano:

> Without mining it would have been impossible to establish an interchange of our products on a large scale, or to achieve the cultural and economic progress attained by our country. The Manila Galleon and the China Ship were the agency for our trade relations with the Orient, climaxed by the Fair of Acapulco, at that time the greatest commercial fair in the world. The fleets of the Old World were constantly putting in at our shores, permitting us to maintain close contact with Europe; and those fleets also had the Fair of Jalapa as their worthy crown.[16]

The situation in Brazil was similar. As soon as the gold of Minas Gerais was discovered, ranches developed in the sertão to supply the miners with meat, oxen, and leather; Antonil's *Cultura e opulência do Brasil* (1711) has a chapter on the abundance of cattle.[17] The great herds which spread throughout the interior developed hardy characteristics, essential if they were to survive in the tropics. In the Northeast, the principal centers were the São Francisco valley where several families with vast estates dominated the scene like feudal lords and the high and open *campos* of Piauí where there were about 400 large *fazendas.* The southern part of Minas Gerais, the Campos dos Goitacases, also became important. One of the most powerful mining entrepreneurs of the period, Manuel Nunes

Viana, *mestre do campo* in the São Francisco region where he had huge holdings, was active in the sale of cattle, and it is interesting to note that Nuno Marques Pereira's *Peregrino da America (Pilgrim of America)* (Lisbon, 1725) was dedicated to him. He undoubtedly subsidized its printing as well as the third volume of Diogo do Couto's *Décadas* (1736), which also bears a dedication to this mining Maecenas.[18]

Taunay clearly understood the stimulating effect of the quest for precious metals on the economy and the westward advance in South America:

> Gold was the real motivation for the definitive possession of the central lands. Had the camp settlements of Bom Jesus do Cuiabá and Guaporé not existed, Alexander de Gusmão [the Brazilian-born diplomat who negotiated for Portugal the Treaty of Madrid with Spain in 1750] would have had a weak basis for invoking *uti possidetis* in favor of setting the boundary of Brazil two thousand kilometers west of the line of Tordesillas. In the seventeenth century the founding of Paranaguá, Curitiba and São Francisco do Sul had already resulted from the search for gold.[19]

As the mining settlements in Goiás (1726) and Mato Grosso (1718) prospered, ranches and small farms grew up around them, and their livestock gave these regions an economic life line. We are apt to forget the significance of livestock in the hinterland; in addition to the use of cattle for fresh meat locally, it was found that salted or jerked beef *(xarque)* could be kept for many months and sent long distances. When one considers that Cuiabá, where the first gold strike in Mato Grosso was made, lies some 800 miles to the northwest of São Paulo, one is impressed that river traffic and mule trails were so speedily developed. Meat and transportation were not the only benefits derived from livestock, however. Hides were equally valuable and the many uses of leather were extraordinary. Capistrano de Abreu spoke of the "Age of Leather" because leather was not only essential for making shoes, boots, saddles, and saddle bags but also for covering doors and beds, for making buckets and cups, and for making jackets and breeches to protect men from the sharp thorns of the *caatinga.*[20]

The opening of the South was due to the Paulistas who had settled Curitiba. The push down the coast toward the Río de la Plata was partly strategic and partly economic. São Pedro de Rio

Grande was founded in 1737, and the great pampa region soon proved to be the best in Brazil for horses, mules, and cattle. The mule, pack animal par excellence, attained special importance, and trails were made over great stretches of the country for these hardy, sure-footed, and patient creatures. So vast were the Brazilian pampas that they became known as the "Continent of Rio Grande."[21]

Since it was necessary to have small centers for salting meat, permanent stock farms *(estâncias)* grew up and a new type of cowhand, the *gaúcho,* came upon the scene. Much like his neighbor, the Spanish gaucho, the southern-Brazilian cowboy developed an independence which set him somewhat apart from his fellow countrymen, and today the Riograndenses are as distinct from their neighbors as are Argentines from Texans in the Spanish- and English-speaking New World.

The chief crops of Portuguese America were sugar, tobacco, cacao, and cotton. Sugar required large plantations and slave labor, and though Brazil never regained the position of world leadership that it held in the sixteenth and seventeenth centuries, sugar exports did well until the last two decades of the eighteenth century. Tobacco, mainly from the Captaincy of Bahia, was valued in Europe, Africa, and Asia (it was used as barter for slaves on the Gold Coast and Angola and to satisfy the smoking tastes of Chinese merchants and potentates)[22] and was second to sugar as an export in the 1770s. Since tobacco could be grown by small farmers and was a government monopoly, it was a blessing to those lacking extensive tracts of land and capital, and it helped to build up a lower-middle class. Cotton production took a great spurt forward in the second half of the eighteenth century, particularly in Pará and Maranhão, with England purchasing the bulk of the crop. Actually Brazil produced more at the time than the North American colonies and maintained this lead until the early nineteenth century. Cacao, also produced primarily in Bahia, was able to compete for some years with cotton as an export and was also consumed locally. Indeed, the domestic market for all crops was a constantly growing one and the great stimulus came from the mines. In considering the history of Brazil one should not ignore the role of domestic trade. Through it the people of the North and South and the hinterland were brought together, and when the time came to sever connections with the mother country, a sense of economic cohesion helped to unite the vastly separated regions.

Notes

1. Alexander von Humboldt, *Essai politique, sur la Royaume de la Nouvelle-Espagne,* book IV, chap. IX.

2. Lucas Alamán, *Historia de México* (Mexico, 1849), vol. I, chap. III.

3. Ibid.

4. Introductory essay to Francisco Hernández (physician-in-chief and historian of the West Indies to Philip II), *Obras completas,* vol. I (Mexico: Universidad Nacional Autónoma de México, 1961).

5. Carlos Pereyra, *Breve historia de América* (1938), part VI.

6. Humboldt, op. cit., book V, chap. XII.

7. José R. Benítez, *Historia gráfica de la Nueva España* (Mexico: Cámara Oficial Española de Comercio, 1929).

8. José de Veitia, *Norte de la contratación de las Indias* (1672) (Buenos Aires: Comisión argentina de Fomento Interamericano, 1945).

9. Thomas Gage, *The English-American: A New Survey of the West Indies* (1648), A. P. Newton, ed. (London: George Routledge & Sons, 1928), pp. 365 and 368–369.

10. The Spanish title of the book is *El lazarillo de ciegos caminantes desde Buenos Ayres, hasta Lima . . .* (Gijón, Spain: Imprenta de la Rovada, 1773). A recent popular Spanish edition is in the Biblioteca de Autores Españoles, vol. CXXII (Madrid, 1959). An English translation by Walter D. Kline was published by the Indiana University Press, Bloomington, Indiana, in 1965. This book gives information about itineraries, vehicles, haulage, merchants, dealers in mules, inns, day's journeys, and post houses for the years 1771 to 1773. Its author was an Asturian born about 1706 in the city of Gijón. He went to the Indies in his youth, and his natural restlessness caused him to travel through almost all Spanish America. "Mexico," he says, "I have traversed from Veracruz to Chiguagua as a merchant, and from Mexico City itself to Guatemala as a traveler." From there he went to the Viceroyalty of Peru, where he was chief magistrate of the provinces of Chilques and Marques. He later lived in Lima, where he married, and traveled quite frequently through the entire South American continent, visiting Buenos Aires in 1749. On a visit to Spain in 1771, he was appointed by the Superintendent of the Royal Postal Revenues as a special "visitor" to reorganize postal service between Buenos Aires and Lima. He began his journey in La Coruña on February 17, 1771, and, with pauses required by his work in Montevideo and Buenos Aires, which he left on Novem-

ber 5, and in Salta, Potosí, Oruro, La Paz, and Cuzco, he concluded his travels on July 6, 1773.

11. Lewis Hanke, *The Imperial City of Potosí* (The Hague: Martinus Nijhoff, 1956), p. 28, citing G. Cobb, "Supply and Transportation for the Potosí Mines, 1545–1640," *Hispanic American Historical Review,* vol. XXIX (1949), p. 27. Also see José Durand, "Vida social de los conquistadores del Perú," *Excelsior* (Mexico), July 10, 1949, and the preliminary study by José J. Real Díaz in the edition of *El lazarillo* published in Madrid in 1959, note 10 *supra.*

12. Arzáns de Orsúa y Vela quoted by William E. Rudolph, "The Lakes of Potosí," *The Geographical Review* (1936), pp. 536–537, and cited in Hanke, op. cit., p. 28.

13. See Charles R. Boxer, *Salvador de Sá and the Struggle for Brazil and Angola, 1602–1686* (London: University of London, Athlone Press, 1952), chap. 3: "The Road to Potosí."

14. Hanke, op. cit., p. 29, citing Helen Douglas-Irvine, "All the Wealth of Potosí," *Pan-American Magazine,* vol. XLIII (1930), p. 160.

15. Quoted in Hanke, op. cit., pp. 29 and 52, n. 91.

16. Gustavo P. Serrano, *La minería y su influencia en el progreso y desarrollo de México* (Mexico, 1951).

17. Antonil, *Cultura e opulência do Brasil,* part IV: "Pela abundancia do gado e courama e outros contratos reaes que se rematão nesta Conquista," pp. 466–487 of the Mansuy edition.

18. Charles Boxer, *The Golden Age of Brazil 1695–1750* (Berkeley: University of California Press, 1962), p. 365.

19. Afonso d'Escragnolle Taunay, *História das bandeiras Paulistas,* 2 vols. (São Paulo, 1954), vol. II, 317–320, in Morse, trans., op. cit., p. 186.

20. João Capistrano de Abreu, *Caminhos antigos e povoamento do Brasil* (Rio de Janeiro, 1930), is a splendid study of the trails, roads, and settlements of Brazil in the colonial period.

21. See José Honorio Rodrigues, *O Continente do Rio Grande* (Rio de Janeiro, 1954) and Moysés Vellinho, *Brazil South: Its Conquest and Settlement* (New York, 1968)

22. Boxer, *Golden Age,* p. 151

A handsome entry distinguishes the royal mint at Potosí, which was erected at the end of the eighteenth century at a time when new mining techniques were being introduced and the Spanish government had developed the coining of local currency. (*Organization of American States.*)

ARTE

DE LOS METALES

EN QVE SE ENSEÑA EL

verdadero beneficio de los de oro, y

plata por açogue.

EL MODO DE FVNDIRLOS TODOS,

y como se han de refinar, y apartar

vnos de otros.

COMPVESTO POR EL LICENCIADO

Albaro Alonso Barba, natural de la villa de Lepe, en la

Andaluzia, Cura en la Imperial de Potosi, de la

Parroquia de S. Bernardo.

CON PRIVILEGIO.

En Madrid. En la Imprenta del Reyno.

Año M. DC. XXXX.

Title page of the first edition of Álvaro Alonso Barba's *Arte de los metales* (1640). In ensuing centuries this important treatise on the amalgamation of gold and silver through the "kettle and cooking" process appeared in dozens of editions and was translated into many languages. (*Engineering Societies Library.*)

A copper-refining operation in Chile was illustrated and described in detail by Peter Schmidtmeyer in *Travels into Chile, over the Andes, in the Years 1820 and 1821* (1824). (*The Hispanic Society of America.*)

The neoclassic Royal College of Mining, erected in Mexico City at the end of the eighteenth century, is the work of Manuel Tolsá. Its elegance and solidity are symbolic of the importance the Mexicans attached to the training of mining experts. (*The University of Texas Library.*)

VIII

Metallurgical Techniques

As mining operations were being established in the sixteenth century, methods for the smelting and processing of metals at first were introduced from Europe. Of particular importance, however, were the techniques of processing silver developed in the New World and further refined here in practice. American progress in this area was so great that it has been the object of admiring commentary from both contemporary metallurgists and modern chemists and engineers.

The first smelting operations by the Royal Exchequer in the New World were carried out with gold in Hispaniola. Later, in 1519, Hernán Martín, one of Cortés's soldiers, operated the first forge and bellows in New Spain. In Veracruz he forged the first iron brought to the mainland from Spain. With other forges also employing Spanish iron, the Spaniards in Tlaxcala fabricated tools, arrows, and other arms. In addition they forged the parts needed for the brigantines that Cortés used on Lake Texcoco to blockade and help capture the Mexican island capital Tenochtitlán, the modern Mexico City. The first metal objects manufactured by the Spaniards with mainland material were some copper lances forged in 1520 on an anvil in Chinantla, near modern Puebla, by Andrés Tavilla, an armorer who came to Mexico with Cortés.[1]

The first working of gold in New Spain was ordered by Cortés from the ransom and booty of Tenochtitlán. But the first actual smelting done by Spaniards from ores they themselves had mined in New Spain was of tin (used with copper to make bronze), as mentioned by Cortés in his Fourth Letter of October 15, 1524, to the Emperor Charles V where he says that "five pieces [of artillery] in all have been finished up to the present: two of them medium-sized culverins, two somewhat smaller, and a serpentine cannon." In Peru, the first operation which melted pure gold into ingots apparently was performed in Tangarara, or Puerto de San Miguel de Piura, by order of Francisco Pizarro before he began his march on Cajamarca.

In both New Spain and Peru, however, smelting was not a satisfactory process, due to the costliness of fuel and the lack of high-grade ore. As a result, after the middle of the sixteenth century, the Spaniards in Mexico introduced a number of inventions and modifications in the methods of amalgamating silver ores with mercury. Despite advances in chemistry, many of their innovative techniques continued to be used for three centuries, almost without modification from a practical and economic point of view.[2]

To be specific, in 1554 in the mines of Pachuca in New Spain, Bartolomé de Medina discovered the so-called patio method of extracting silver from its ore by the use of mercury. A patent covering the jurisdiction of the Audiencia of Mexico was issued by the Viceroy Luis de Velasco in 1555 to enable Medina to collect royalties from those miners using his refining system.[3] Sufficient mercury was available in Mexico and the success of the patio process there was such that by 1562, in Zacatecas alone, thirty-five establishments were using this method, which permitted the processing of low-grade ores not suitable for smelting.

In its original form, amalgamation by the patio method was a cold process which consisted of four essential steps: (1) crushing the silver ores with mallets or mortars and then, usually, completing the pulverization with grinding mills or sledges; (2) treading the so-called *repasos* (the ground and moistened mass) by men or horses, mixing in salt, mercury, and, generally, roasted copper pyrites *(magistral),* and then spreading it in large cakes *(tortas)* on the floor of an open or roofed courtyard (the patio); (3) washing the material with water in tubs equipped with beaters to separate the silver amalgam; and (4) distilling or retorting the amalgam, generally in receptacles known as *capellinas,* to free the silver and recover part of the mercury. The whole process required three weeks or longer for completion.

A variation of the amalgamation process, and apparently the first method invented by Bartolomé de Medina, was another cold process using troughs (called *canoas* in Mexico). Bargalló has established that, properly speaking, this was an antecedent of the later-improved, more widely used patio method which Medina may not have lived to see.[4] In the method in general use in New Spain, all the repasos were treated by trampling the material in wooden troughs. But in Pachuca, less than twenty years after Medina's invention of the extraction method, miners already were using *estufas* (ovens) in a process which mixed the material in the form of large balls placed in piles on the inner floor of the closed oven.

Although mercury was available in Mexico, there was some question about whether it could be found in Peru. A Portuguese merchant and poet by the name of Enrique Garcês, who arrived in Lima in 1547, had been intrigued for some time by the fact that Indian women painted their faces with cinnabar, and he remembered that this red powder was usually found near mercury. When, on a visit to Mexico in 1558, he learned that Bartolomé de Medina's patio process required mercury in treating silver ore, he saw the possibilities of exploiting the process in Peru. Returning there the next year with a Mexican miner, Pedro de Contreras, Garcés compared his Peruvian finds with Spanish mercury from Almadén, and, after demonstrating the patio method to the Viceroy, obtained a twelve-year monopoly in the area to develop silver production with mercury. Garcés was the first to work the Huancavelica lode, and he also made some technical improvements in mining techniques. His scientific studies resulted in the invention of a more efficient mercury smelter. He built a heating system composed of a central round furnace similar to the cupola furnaces used to refine silver, and two lateral kilns *(buitrones)* of the kind then used in Huancavelica. All three connecting units were filled with kettles containing ore, but only the two side kilns were fired. Garcés theorized that the heat emitted by these lateral units would irradiate the central furnace, supplying enough energy for smelting through reverberation. His first design, tested in July 1581, was unsuccessful. However, after repeated trials and several minor modifications in the round furnace, he was able to prove, in January 1582, that the furnace arrangement he had devised achieved considerable fuel savings and higher product yields. After acquiring a sizable fortune, Garcés went back to the Peninsula and, being interested in translating, brought out a Spanish version of Camões *Lusiads* at Madrid in 1591. Cervantes praised the Portuguese poet-entrepreneur warmly in *La Galatea*.[5]

Since mercury was the basis for the extraction of silver by the amalgamation method or its various adaptations, the discovery of mercury mines in Peru was extraordinarily important. During the years 1571 and 1572, Pedro Fernández de Velasco adapted Garcés's patio process in the form of the trough method to the ores and climatic conditions of the Peruvian-Bolivian plateau in general. This innovation gave new stimulus to the mines of Potosí, which had begun to decline.

The trough procedure used in mines in Upper Peru and other mines in Old Peru was at first a cold process. After a few years, however, the operators began to heat the troughs, which were made of stone or brick, and thus converted them into buitrones. As Bargalló has established, both these Peruvian procedures originated in Mexico: the cold process in central Mexico and the hot process in Pachuca, where the estufa method, which inspired it, had certainly been used before 1575. Finally in 1590 in Upper Peru, Alvaro Alonso Barba invented his famous method of processing ore by a variation of amalgamation called "kettle and cooking."[6] In extraction by this method, amalgamation was achieved by heating the finely pulverized ore, along with water and mercury, to the boiling point in kettles made of pure copper and equipped with beaters for stirring the mass. The mixture had to boil throughout the process; the fire was stoked and water was added as necessary. This process, as Barba has said, reduced to a few hours the days required by the ordinary cold process.

In his *Essai politique*, Humboldt makes the following valuable observation: "In 1609 Alonso Barba proposed amalgamation by the use of heat or by cooking in copper vats: this procedure is called 'kettle and cooking'; and it is the one proposed by Mr. de Born in 1786."[7] And Bargalló correctly comments: "If Central European metallurgists had realized that they owed anything to methods invented in the Indies, extraction by amalgamation would certainly have become established in Chemnitz or Freiberg [both in Saxony] as a result of Barba's invention [1609 to 1616], or perhaps sooner, for Medina's discovery dates from 1555."[8] The famous metallurgist Friedrich Traugott Sonneschmid, who stayed in New Spain for more than ten years at the end of the eighteenth century in the employment of the Spanish government, refers to this contempt for Spanish American metallurgical methods: "It would be well, however, if recognition [of the patio method] were to spread to Europe in order to do justice to this most excellent method which European opinion has treated with considerable scorn. . . . I have no hesitation in declaring that in ten years of work I have not been able to find either Born's process or any other preferable to the patio method."[9] This unjustified lack of confidence in Spanish

American methods by Europeans became widespread despite the very great reputation attained by Barba after his system became known through the publication of his *El Arte de los metales* in the seventeenth century.

The Huancavelica[10] and other lodes, however, did not last indefinitely, and Cañete wrote in 1787:

> These same mines supplied all the mining establishments in Peru with a prodigious quantity of mercury for more than two centuries, until the year 1752, when the supply began to decline; thenceforward mercury was also brought to Peru from Almadén, which formerly had supplied only the mines of Mexico with five or six thousand quintals a year, after which time the output was increased to as much as eighteen thousand quintals annually. . . . However, since the production of both mines was not sufficient to supply the two kingdoms of Peru and Mexico, our Court has contracted with the Imperial Chamber of Germany to ship annually to us in Cádiz a sufficient quantity of mercury to supply the Royal Storehouses of those kingdoms with the necessary amounts of this ingredient. This has confirmed the wise prediction of the mineralogist Bowles, that the time would come when the mines of Almadén as well as those of Huancavelica would be inadequate, and advising that it would be most recommendable to seek in good time for other mines with which to supply ourselves, since in all the world no other mercury mines are known except those of Carinthia in Germany, Huancavelica in Peru, Almadén in Spain, and Triulí in Hungary, apart from Hydria, which is the chief one.
>
> Hence the conjecture made by Señor Solórzano, that if the mines of Huancavelica were well husbanded they could supply mercury ore to all the Indies, is seen to be mistaken; for, his calculations being based on past times, recent times have seen the decline of that district, to the point that we are obliged to bring mercury from Germany, as has been said, and as Señor Solórzano himself testifies had been done on other occasions.[11]

According to Francisco Xavier de Gamboa, the eminent eighteenth-century Mexican jurist, the principal mercury mines in New Spain were those of Chilapa, explored in August 1676 by Gonzalo Suárez de San Martín; those of Sierra de Pinos, in New Galicia; two new mines in Cerro del Carro and Cerro de Picacho, discovered in 1740 by Don Felipe Cayetano de Medina and Don Gregorio Olloqui; and those of Temascaltepec, discovered in 1743.[12]

Mercury was so indispensable in the production of silver that, for purposes of registration and to guard against contraband or fraudulent production, the amount of mercury used by a mine served as a basis for calculating and collecting the Royal Fifth. Hence the Guild of Mercury Producers was formed in Potosí in order to deal with the authorities in the procurement, shipment, distribution, price, and other regulatory matters having to do with mercury. Gamboa says: "In no other thing are the power and supreme prerogative of His Majesty more noteworthy in the mines, than in regard to those of mercury, and the sale and distribution of this ingredient."[13] He also comments that "sometimes it was brought from Peru, from the celebrated mines of Huancavelica; at other times from Spain, from the mines of Almadén; at others from Germany; and at others from China by way of Acapulco." All the handling of mercury—its licensing, traffic, sale, price, and distribution—was reserved to the Crown. One of the Laws of the Indies states: "Only by permission of the Royal Exchequer may mercury be dealt in, on pain of confiscation, and its sale by merchants and miners is prohibited, even though it be part of what has been distributed to them by the Royal Exchequer."[14]

There were no great technological discoveries in Brazilian mining for a number of reasons. The distance of the gold fields from the coast and the vastness of the territory meant that frontier conditions would prevail for some time. The Brazilian historian Sergio Buarque de Holanda points out that one of the plates of Agricola's *De re metallica* (1556) could almost serve to illustrate mining practice in the early days of Minas Gerais, the small iron hoe being similar to the Portuguese miner's pick, or *almocafre,* the vessel or bucket being clearly an ancestor of the *carumbé,* and the potbellied oval bowl in which to wash the gold-bearing earth being similar to the Brazilian *bateia.*[15] However, foreign specialists were sent over by the Crown as early as 1700, and by 1707 we hear of artificial reservoirs and dams being built far up on hillsides to take advantage of the rainy season and furnish water power to wash down rocks and masses of earth which were then channeled through sluices and troughs so that the gold dust and nuggets could be separated from them. According to Claudio Manuel da Costa, it was a priest with the humorous nickname *Bonina Suave* (Gentle Daisy) who was responsible for developing a waterwheel in 1711 to sort stones from the precious metal *(esvaziamento das catas)*. Manuel

Pontes apparently improved the waterwheel around 1725, for in that year he obtained a patent to make such machines. The working of mountain hillsides began in 1721 when Governor Lourenço de Almeida issued a proclamation allowing mineshafts to be dug on the Morro de Mata Cavalos providing they were at least 8½ meters apart. Crushing machines were also installed, for instance, on land belonging to Padre Manuel Gomes Neto at Taquaraçu.[16] Underground workings did not develop because of the nature of the ore. Instead, hydraulic works in the form of damming, sluicing, and sifting for separating gravel from the deposits continued throughout the eighteenth century. Before the days of hydraulic nozzles and power-driven dredges, efficiency, in modern terms, was not possible.

Mining for diamonds was not too different a process. The bateia, however, gave way to sieves, conduits, sluices, and troughs. As with gold, huge piles of gravel were dug from river beds during the dry months, and when the rainy season came (October to April), streams and water from dammed-up reservoirs were used to wash the diamonds free.[17]

With the arrival of Dom João VI in Rio de Janeiro (1807), all the country's ports were opened to the world, and skilled European experts and Brazilians trained abroad in mining techniques were given the task of modernizing the industry. They found, however, that it was not easy to improve significantly on the metallurgical methods of their predecessors.

Notes

1. See Modesto Bargalló, *La química inorgánica y el beneficio de los metales en el México prehispánico y colonial* (Mexico City: Facultad de Química de la Universidad Nacional Autónoma de México, 1966), p. 71.

2. St. Clair Duport, *De la production des métaux précieux au Mexique* (Paris, 1843), quoted by Bargalló, *La minería, y la metalurgia en la América Española durante la época colonial* (Mexico, 1955), p. 129. According to Bargalló, "A number of intelligent miners, workers, doctors, lawyers, priests, and soldiers presented and practiced a great many innovations in the amalgamation process, some of them of considerable value: Alonso Martínez de Leiva, 1560, of New Spain; the cleric Antonio Boteller, in 1562; the miner Juan Capellín of Tasco [New Spain] in 1576; Gaspar Ortiz of Potosí, with his treatment of roasted ores and residues with brine, in 1586; Carlos and Juan Andrea Corzo, miners of Potosí with their extraction process using iron filings, 1587 [which reduced the loss of quicksilver]; the *bachiller* and cleric Garci Sánchez, with his

methods based on sulphur, 1588, and utilizing slag [*escorias o tierras*], also attributed to Domingo Gallego in Peru; Juan Fernández Montano of Potosí, 1588, with the use of copper sulphates and brine; Francisco Pacheco and Pedro Poblete, with their amalgamation process involving as a first step the burning of cuprous silver ores with a ferrous admixture, 1602, in the kingdom of Peru; a Franciscan monk, with the *tintín* process; Captain Pedro Mendoza Meléndez, Pedro García de Tapia, and Dr. Berrio de Montalbo, in 1643, with their barrel process [*beneficio de la barrilla*] in New Spain; Juan Corrosegarra of Peru, in 1678, with his method of treating pellets [or lumps] of silver with quicksilver; Lorenzo de la Torre, in Peru in 1738, with his extraction process using peroxides of iron [*colpa*]; the Presbyter Juan de Ordóñez, in Pachuca in 1758, with a new oven method [ovens had been in use in New Spain since the end of the sixteenth century]; the lawyer José Garcés y Eguía, who perfected the washing procedure and the "patio" method in general, 1799. Friedrich Sonneschmidt, a German mineralogist employed in New Spain, should also be mentioned, for he improved some aspects of the patio method at the end of the eighteenth century." (*La mineria,* pp. 112–113.)

3. In 1556 a revised patent was granted to Bartolomé de Medina and to Gaspar Loman, a German, to exploit this process. The new patent was conditional, and in 1560 it was extended only to Bartolomé de Medina since Loman's method of handling ores within the patio did not work successfully. It now appears certain that discovery of the patio process may be attributed to Bartolomé de Medina alone. This has been investigated by Luis Muro of the Colegio de México and also by Bargalló, according to his book *La amalgamación de plata en Hispanoamérica colonial,* to be published soon. Also see the excellent treatise by Alan Probert, "Bartolomé de Medina: The Patio Process and the Sixteenth Century Silver Crisis," *Journal of the West,* vol. VIII, no. 1 (January 1969), pp. 90–124. Probert claims that Medina's affidavit of discovery, dated December 29, 1555, should have been dated 1554 because at the time all days after Christmas were regarded as belonging to the next year.

4. Bargalló, *La química inorgánica,* pp. 89–90.

5. Guillermo Lohmann Vîllena, "Enrique Garcés, descubridor del mercurio en el Peru, poeta y arbitrista," *Anuario de estudios americanos,* vol. V (Seville, 1948), pp. 439–482. Luis Monguió, "Los versos del perulero Enrique Garcés y sus amigos (1591)," in *University of California Publications in Modern Philology,* vol. LVIII (1960), no. I. The publication read *Los Lusiadas de Luys de Camões, traduzidos de Portugues en Castellano por Henrique Garcés: Dirigidos a Philippo Monarcha primero de las Españas, y de las Indias. En Madrid, Impreso con licencia en casa de Guillermo Drouy impressor de libros. Año de 1591* . For all above see Hanke "The Portuguese in Spanish America, with Special Reference to the Villa Imperial de Potosí," *Revista de Historia de América,* vol. 51 (June 1961), pp. 18–20.

6. According to the entry in the baptismal register, Alvaro Alonso Barba was born in Lepe, province of Huelva, in Andalusia, on November 15, 1569. He took holy orders and went to the Indies. As early as 1609 we find him in Upper Peru (modern Bolivia) where he served in several parishes. It was while he was at San Bernardo in Potosí that he wrote his book *El Arte de los metales*. Barba had dedicated years to the examination of silver-bearing sites and soils, as well as to a careful and exhaustive study of the extraction of metals. Padre Alonso Barba returned to Spain in 1657 and died in Seville at the age of ninety-three. At the time of his death he was nostalgically preparing to return to the district of the Audiencia of Charcas, the chief setting for his long and useful life which he dedicated to the art and science of metallurgy and the very doubtful care of the souls of those who extracted metal from the land.

7. Alexander von Humboldt, *Essai politique sur la Royaume de la Nouvelle-Espagne,* book IV, chap. XI.

8. Bargalló, *La minería,* p. 113.

9. Federico (Friedrich Traugott) Sonneschmid, *Tratado de amalgamación* (Mexico, 1805), prologue and chap. 26, cited in Bargalló, *La minería*.

10. Ibid. The classic study of Huancavelica in English is by Arthur P. Whitaker, *The Huancavelica Mercury Mine,* Harvard Historical Monographs, no. 16 (Cambridge, Mass.: Harvard University Press, 1941).

11. Juan de Solórzano, *Política indiana* (1647), book VI, quoted in Pedro Vincente Cañete y Domínguez, *Guía histórica, geográfica, física, política, civil y legal del Gobierno e Intendencia de la Provincia de Potosí* (1787). A modern edition of Cañete was published by Editorial Potosí (Potosí) in 1952.

12. Francisco Xavier Gamboa, *Comentarios a las Ordenanzas de Minas* (Madrid: Oficina de Joaquin Ibarra, 1761), p. 25.

13. Ibid.

14. Ibid., chap. 2, par. 38.

15. Sergio Buarque de Holanda, *História geral da civilização Brasileira: A época colonial,* tomo I, vol. 2 (São Paulo, 1960), pp. 274–275.

16. Ibid., p. 275.

17. In the Palácio da Ega in Lisbon there is a 1750 painting showing a diamond-working center in the Sêrro do Frio. The covered, wooden conduit, bucket wheel, and washing troughs clearly indicate that the bateia technique handled by individual workers in a stream was no longer practiced by the Crown contractors.

IX

Legal Edicts and Mining Institutions

From the beginning of the occupation of the Indies, the mining industry was supremely important by virtue of its significance in the economy of the new territories and the King's finances. Hence it was only natural that a whole fiscal, juridical, and administrative structure should be established. This structure administered the legal regulations pertaining to rights of acquisition and exploitation of mines, the study and resolution of the litigations which arose, and the imposition and administration of the Crown's shares in the profits of mining. Therefore, in addition to the political offices set up to govern the colonies—Adelantados, Viceroys, Captains-General, Governors, Ayuntamientos (city councils)—the Crown also established other institutions—Audiencias, tribunals, loan banks, silver purchase banks—to deal specifically with mining activities either full-time or as part of their many duties, and enacted a special body of mining regulations. The laws in force in Spain, from the medieval *Siete Partidas* of Alfonso X (the "Wise") to the *Ordenanzas de Castilla* of the Catholic Sovereigns Ferdinand and Isabel, were applied in the newly discovered lands and gradually adapted to the circumstances and needs of the New World.

Before the discovery of America, the beginnings of a juridical system concerning mining had begun to evolve in Spain.[1] These enactments were important because they established the future position of the sovereign vis-à-vis ownership of the subsoil. Under the civil law[2] mine titles were not absolute, and the Monarch—as well as the owner of the surface land—could claim a tenth of the value of the metal mined.[3] Alfonso X severed mineral rights from land grants—a principle still observed today in Latin American nations—and made the collection of the Crown's portion part of his sovereign prerogative.[4] Alfonso XI made all minerals not specifically granted already part of the Royal Patrimony,[5] while Juan I decreed that unlicensed prospecting for mines on open lands depended upon payment of two-thirds of the net profits to the Crown. Within the confines of a land grant, however, this right to prospect freely was extended only to the proprietor of the surface; he could bar third parties from prospecting.[6] Hence, mines were a royal prerogative reserved exclusively for the King. But this reservation did not mean that private persons could not exploit them in usufruct, i.e., with the payment of a tax or rental fee. This fee varied a great deal according to the period, types of metal, and economic results of mining exploitation. In fact, as a practical matter, all mines were worked under various concessions and contracts by private individuals, not Crown servants or employees.

Columbus's discovery of a new land with unforeseeable opportunities and great potential mineral wealth caught the Catholic Sovereigns unprepared. Having set their claims and position in this New World by Papal Bull and international treaty, the Spaniards turned to exploiting their dominion. The Indies belonged to the Crown of Castile as a personal possession of the Monarch, and according to legal doctrine, property rights existed only as an indulgence of the Crown. In actuality, however, the Crown had to concede extensive rights to many people and groups in order to foster exploitation of the land, and much of the extension of Spanish power in the Indies was accomplished through private enterprise.

The first pronouncements concerning mining in the New World carried from Spain the idea of Juan I that any discoverer of gold could keep one-third, but that allowance was soon increased as a practical matter to one-half and then four-fifths. While numerous regulations and orders were issued, they appear to have been set on an ad hoc basis, and except for an unwavering insistence on the Royal Patrimony over minerals (the subsoil, in modern parlance), there were several changes in policies. However, some general rules may be traced to royal actions dating as far back as 1530. Immediately after the discovery of the Indies, the Crown reserved for itself the benefits of the mines with the exception of those for whose exploitation it would extend a special concession. In 1504, the general rule was established that mines could be searched for and worked without securing previous permission from the Crown on condition that successful prospectors would have to pay the Crown 20 percent (or less, in some cases) of the net returns in metal. The payment of the Royal Fifth was not regarded as a tax, but rather as a royalty. Shortly afterward, there appeared the doctrine that exceptionally rich mines were reserved for the Crown, although later the discoverers of such mines were allowed under certain conditions to exploit them. In addition, during the sixteenth century, a portion of all mines discovered was reserved for the Crown. This practice disappeared in Mexico by the end of the century although it was recognized in Peru until the end of the colonial era.[7]

The first mining law written in the Indies was issued in 1532 by the Audiencia of Mexico City. This law was short-lived, however, being superseded by a new one issued in 1536 by the Viceroy Antonio de Mendoza and designed explicitly for use in Mexico. Mendoza's Law, regulating and protecting mine labor, was supplemented in 1539 and 1543 by others which governed the exploration, registry, and exploitation of mines. Because of the confusion that arose over the application of these enactments, on January 14, 1550, Mendoza promulgated the first American inclusive mining code.[8]

Mendoza's Code did not touch upon the topics of mines and the Royal Patrimony, nor the status of Indian rights, because they were already defined by other legislation. But it did prescribe the conditions under which anyone—Spaniards as well as Indians, who were also regarded as free vassals of the Crown—could discover, claim, acquire proprietorship, and work a mine. Rules concerning registry and the marking of boundaries were quite strict, leaving the impression that "claim jumping" was widely practiced. Long sections of the code prescribed working conditions, which particularly concerned Indians, in order to ensure their health and security.

In the case of Peru, the first law concerning mining in that Viceroyalty was issued in 1561. However, it soon proved inadequate, and a comprehensive code embodying regulations on the same matters as Mendoza's Mexican Code was issued by the Viceroy Francisco de Toledo in 1574 specifically to regulate the mining industry in Peru.[9]

The Codes of Mendoza and Toledo eventually were afforded different statuses in the view of the Crown. Philip II formally approved Toledo's Code in 1589,[10] and, with various additions, it remained the basic mining law for Spanish South America until 1785–1786 when the newly enacted Mining Ordinances for New Spain of 1783 were ordered to be applied with the necessary adaptations to Peru. The exact status of Mendoza's Code between 1584 and 1602 is difficult to determine because of the paucity of documents. An order to enforce the laws of Castile in New Spain, with particular regard to the mining ordinances, may have antedated Philip III's order of November 26, 1602, when he decreed that the Audiencias "send us a particular account of what laws of mines are not enforced in each province, and why, and also the reasons they may have for ordering those to be observed which they may deem necessary."[11] Arthur S. Aiton notes that in 1577 the Audiencia of Mexico City had advised the King that there were variations between the royal ordinances of Castile and those of Mendoza and that the latter was determining law suits. Twelve years later, in 1589, after the promulgation of a new Spanish mining code in 1584, the same Audiencia asked the Council of the Indies to reconcile the new royal code with actual practice which was based on Mendoza's Code. No answer to this request has been found, but neither is there evidence of any drastic change of policy by the Crown which would injure Mexican interests.[12] Thus, the exact date on which the Code of 1584 became effective in New Spain is still a moot question. The disappearance of Mendoza's legislation from commentaries and histories would indicate that the influence of this code was confined in time and area of application. Hence, the development of mining jurisprudence in New Spain for some 200 years was chiefly based on legislation issued in Spain.

Spanish mining law had remained static between 1387 and 1559, when the Crown promulgated a series of laws setting forth minute bureaucratic regulations governing the industry. The concepts of regalism in the control of the mining industry and the promulgation of a complete code of regulations first developed in the New World through the activities of the royal agencies and the royal officials appointed to rule those domains. When royal power in the Iberian Peninsula became strong enough to hobble the pretensions of the nobility, the concepts developed in the Indies were at last made a part of Castilian law. Aiton has noted that on July 29, 1578, Philip II requested a copy of Viceroy Mendoza's mine ordinances and that they were duly forwarded to the Council of the Indies.[13]

On January 10, 1559, Princess Doña Juana, in the name of Philip II, issued a decree which reiterated the royal prerogatives in mining laid down in the fourteenth century and prepared the foundation for the comprehensive mine codes which were to emerge over the next twenty-five years.[14] Although many of the principles of the decree were already in force in laws relating to the Indies, the decree did lay the foundation for a number of continuing principles and practices in Spanish American mining law. These included:

1. The principle of Royal Patrimony in the subsoil with the implication that the mining industry was of interest basically to the Treasury before any other government organ
2. The right of free prospecting in public and private lands without prior authorization or the proprietor's consent
3. The necessity to register a claim in order to secure recognition of discoverers', owners', and operators' rights
4. The right of the discoverer or the miner to exploit his mine and refine his ore without further specific authorization once the mine was duly registered
5. The principle that rights to subsoil minerals would be legally recognized only after the deposit was actually discovered and duly registered
6. The principle that a mine must be worked if its owner's title was to remain valid

Four years later, on March 18, 1563, Philip II issued a series of seventy-eight ordinances which supplemented and partially modified the edict of 1559.[15] They are generally known as *The Old Ordinances of Mines (Ordinanzas del Antiguo Cuaderno),* the title applied to them by Gamboa in his *Commentaries.* Several provisions of the 1559 decree were repeated with almost no variation, others were clarified, and others were modified only in part. The Royal Patrimony over mines was not mentioned, since the previous edict was regarded as being incorporated into the ordinances.[16] These articles introduced the principle of variations in the King's portion based upon the kind of metal (base as well as precious metals were enumerated) and the nature and grade of the ore. In an article that expanded and made more specific the general right of open prospecting throughout the country, foreigners were also permitted to undertake the tasks of prospecting for and operating mines.

On August 22, 1584, Philip II issued a new mining code entitled *The Ordinances of Mines,* usually referred to as the "New Ordinances" or "New Code of Mines" *(Ordenanzas del Nuevo Cuaderno),* following the usage of Gamboa.[17] In Article 1 the new code revoked the edicts and orders of 1559 and 1563 and all other laws, patents of privileges, and customs only "so far as they are in opposition to the provisions of this Law. . . ." The only exception to this was the blanket incorporation of all mines of gold, silver, and mercury into the Royal Patrimony.[18] The article then stated that "said mines be worked, and reduced, and that all suits and disputes be decided and determined" under this new law.

In Article 2, Philip introduced an innovation: Spanish subjects and natives as well as foreigners "who may work or discover any mines of silver, already discovered or to be discovered" were to have them as "their property, in possession and ownership, and that they may do with them as with anything their own," providing they paid the duties imposed by the code. This article constitutes the first concrete statement in Spain of the previously implied concept of dual ownership of the wealth of the subsoil. While the Crown maintained its ultimate patrimony in the soil, it extended certain wide-ranging privileges to the discoverers and exploiters of silver mines. Although they were not extended *fee simple* rights, those who discovered mines were given an almost free hand in making business arrangements based upon the new security of tenure. While the miner remained, in many aspects, a royal agent whose work contributed to the fisc, he was given great leeway in conducting his business. A key point must be noted, however: Even this limited right to "possession and ownership" was confined to those who discovered, registered, and worked mines.

The remainder of the ordinances repeated the principles set forth in the earlier decrees, although usually in greater detail. Continuing problems of the growing mining industry in Spain, together with the experience of the Germans, contributed to the length and detail of the instructions covering registry, boundaries, working practices, business procedures, the adjudication of legal disputes, and assurances for the collection and remittance of the King's revenues.

The Code of 1584 remained in force in Spain without any basic changes until the nineteenth century. In the New World, perhaps the greatest tribute paid the New Code of Mines is the fact that for 199 years no plan or set of rules was recommended to replace it and only a few modifications were made in its operation. Changing conditions in the Indies finally necessitated the framing of a successor to the New Code in 1783 for specific application to New Spain and later to Peru, but that code was based in great part upon the 1584 law. The Codes of 1584 and 1783 finally succumbed only to the changing legal and economic philosophies of the nineteenth century.

During the period between 1584 and 1783 two classic studies were made of the mining industry and its relation to the Crown. The first, Juan de Solórzano y Pereyra's *Política indiana (Policy For The Indies),* 1647, surveyed the entire field of relations between the Crown and the Indies and devoted one chapter to the problems of mining. Solórzano was a member of the Audiencia of Lima and later served on the Council of the Indies in Madrid. His work on Spanish colonial law is often compared to Blackstone's *Commentaries* in its influence. Not only was his analysis a concise statement of the status of existing mining law, but his explanation of origins, developments, and problems are invaluable. Gamboa devoted his entire lengthy *Comentarios a las ordinanzas de minería (Commentaries on the Mining Ordinances),* 1761, to a summary and analysis of the legal history of the New Code of Mines, its technical applications, and an exposition of the technical mining and processing problems of the far-flung Indies. A great many of Gamboa's constructive suggestions were adopted in the revision of the mining code issued in 1783, and his book continued to be regarded in Mexico as the authoritative commentary on the legal aspects of mining problems. Until the end of the nineteenth century, Mexican jurisprudents cited Gamboa as the basic authority on the old codes and interpreted the Code of 1783—which remained in force in Mexico until 1884—in the light of his analyses.

During the last half of the eighteenth century a spreading paralysis afflicted the mining industry of Mexico. New discoveries were no longer being reported, and production declined. Reorganization of the industry seemed imperative. To aid this "fountain of wealth," a great number of suggestions were offered for legal reforms and government action. Gamboa devoted several chapters of his *Commentaries* of 1761 to outlining the financial aids and administrative changes he felt were necessary to save the industry. Inspired by a *Representación* made in the name of the miners of New Spain by a Spaniard, Juan Lucas de Lassaga, and a Creole, Joaquín de Velásquez Cárdenas y León, José de Gálvez, who had resided in New Spain as Inspector General between 1765 and 1772 and now was Minister to the Indies, undertook negotiations with Charles III for a new mining law.[19] After a lengthy study of the problem, the King, on May 26, 1783, approved a document entitled *Mining Ordinances of New Spain.*[20] This code was destined to endure for 101 years before the Mexican Republic repealed it in its entirety.

To help the mining industry recover, the ordinances authorized a miners' bank and a school of mining engineering. The school was greeted with enthusiasm, for it would provide a source for the necessary trained personnel. The ordinances also set up a new series of mining districts, with new officials to be elected by the mine owners, established an extrajudicial system of judging mine disputes and regulating the workings of the industry, and enumerated a series of special privileges for members of the Mining Guild. In addition, the ordinances contained long and detailed descriptions of the approved methods of exploiting mines based on the New Code of 1584 and Gamboa's *Commentaries*.

In 1785–1786 the ordinances for New Spain, with the necessary modifications, were ordered into effect in Peru. On the basis of these regulations and the jurisprudence established in the course of years, Don Pedro Vicente Cañete in 1794 drew up a treatise entitled *Ordenanzas de las minas del Perú y demas provincias del Río de la Plata (Mining Ordinances of Peru and also of the Provinces of Río de la Plata)* at the order of the Governor of Potosí, Don Francisco de Paula Sanz.

The Ordinances of 1783 marked the final stage in the development of regalism in mining. Only one basic change by Spain—or independent Mexico—was made in the application of this concept: the Crown decided in 1789 to encourage the exploitation of coal mines by placing them in a special category. Several decrees concerning this matter were issued in the next four years. The emancipation of New Spain from Spanish rule did not entail any changes in the ordinance: unless specifically repealed or contradicted by Mexican legislation, Spanish law remained in effect in the new

Republic. Mexico did not repeal the ordinances in their entirety until the passage of the Mining Code of 1884.

Administration of the Royal Patrimony in the Indies was a weighty and complicated matter since it included the vacant lands, royal domains, and mines. The amounts of money collected for the Crown's account were enormous. When the first bonanzas were being opened, collection of this portion of the Royal Treasury receipts had been made the responsibility of the Governors. Then it had been transferred to the jurisdiction of the Supreme Council of the Indies. As the responsibility grew and the problems of administration and audit and the maintenance of probity became more difficult, a new organizational structure was erected. A Tribunal of Accounts was set up in the sixteenth century to be in charge of financial operations. In turn, a Junta Superior del Real Hacienda (Higher Committee of the Royal Treasury) composed of the Viceroy or Governor, Treasury officials, a senior judge, the *fiscal* of the Audiencia, and the senior auditor of the Tribunal of Accounts was established to set treasury policy and check on warrants for actual disbursements. The chests of the depositories were secured with three or four locks with an equal number of officials holding one key each: hence all had to be present for the depository to be opened. In Potosí the Royal Depository to hold the treasure was built in 1598 at a cost of 15,000 gold pesos. A reform was carried out in the management of the Royal Depositories in the eighteenth century when they were made dependent on the Intendant for each province. The Intendant also had jurisdiction over lawsuits in which the Royal Treasury might have an interest.

After the conquest, says Cañete, in order to avoid trade "by exchange and barter of some things for others, there were introduced dealings in silver bars, small gold bricks, and even in gold dust." But this type of medium of exchange was subject to so many variations in value, such as the degree of fineness or purity of the bars, as well as their weight and volume, that the authorities were obliged to test these metals in order to determine their genuineness or purity. They then had to divide them into fractions by imprinting their weight (*peso,* hence peso the monetary unit) and value.

It therefore became necessary to mint coins of a genuineness and value assigned to them by the state. At the prompting of Bishop Zumárraga and other royal officials, a cedula was issued on May 11, 1535, authorizing the construction of mints in New Spain and on the island of Hispaniola and declaring that the mints follow the same standards set by those of Castile.[21] While the mint on Hispaniola was constructed but not used, the Mexico City mint was set up immediately by New Spain's first Viceroy, Antonio de Mendoza, and by 1536 it began to coin silver and copper pieces. The mint at Lima began operations in 1568 and continued to operate sporadically, but since it was far from the source of bullion, another mint was opened in 1572 in Potosí.[22] The Peruvian mints worked exclusively with silver, and the Mexican house had to discontinue its production of copper coins because the Indians despised them and would not use them—one report had them throwing the coins into Lake Texcoco! For a long time after the conquest, in fact, the Mexican Indians continued to conduct their small transactions with their ancient coinage, cacao beans! The minting of gold, on the other hand, was prohibited in the Americas until late in the seventeenth century. But in 1683, the Viceroy, the Duke of Palata, finally was ordered by the Crown to construct a mint at Cuzco especially for the coinage of gold. In the eighteenth century, mints were opened at Bogotá and Popayán to coin the gold output of the Viceroyalty of New Granada, which in the late eighteenth and early nineteenth centuries was probably the world's leading gold producer.

Bailey W. Diffie has summed up well the influence the establishment of a mint had upon a remote sector of the Spanish Empire marked by a sluggish and backward economy:

> Chilean production of metals in the colonial period did not compare in value with that of Peru and Bolivia, but nevertheless the mining industry was important enough to influence Chile's economy. By the end of the eighteenth century, her mining activities were expanding.
>
> A mint was established in Santiago de Chile in 1749 and began to function early in 1750 coining both gold and silver. . . . The lack of specie, which had been felt before, was remedied, and Chilean commerce was carried on with the new currency. Silver and gold mining received a stimulus since miners now had a market for their precious metal without having to send it to Peru. . . .
>
> The increase of population, trade, and wealth of Chile was due in large part to the advantages of the mint and mining. . . .[23]

Sometimes mills for processing ore were located at great distances from the cities where mints existed. Because of the difficulties and dangers of the roads, it was both inconvenient and financially damaging to those miners to ship out their gold and silver bars and then wait until they received their value in new coins. This delay even affected labor at the mines, for sometimes there was not enough money on hand to pay the miners' wages. At the same time, the mints often would not have enough metal with which to operate. The result was a serious and widespread detriment to a vigorous economic life. To make matters worse, certain merchants speculated with silver bullion to take advantage of this situation. Their operations made for conditions that were onerous for the miners and prejudicial to the Royal Treasury.

To remedy this state of affairs, the Silver Purchase Banks (called *Bancos de Rescate* or *Bancos de Plata*) were organized.[24] Their basic function was to make certain that the mints always had enough coined money to pay the miner a fair price for his metal as soon as he brought it in. The first bank of this type was organized in Potosí in 1752 by the Mercury Producers' Guild and later passed to the control of the state under the name of Royal Redemption Bank of San Carlos.[25] A royal cedula in 1776 created Silver Purchase Banks in New Spain. The first was established in San Luis Potosí, but after 1791 the same regulations for redemption of silver were applied in the mining establishments of Zacatecas, Pachuca, Sombrerete, Rosario, Zimapan, Chihuahua, Durango, and Guanajuato.

In contrast to Spanish America, Brazil developed a mining industry only late in the colonial period—after 1695—and the evolution of her mining law and government institutions to handle the precious metals was naturally delayed. Since Portugal had no mining industry to speak of, and the first two centuries of the occupation of Brazil yielded only minor discoveries of metals, the mining jurisprudence of Brazil had neither the background nor the complexity of Spanish American mining law. Furthermore, since Brazilian precious metal mining was almost all the placer type, it did not require many of the details concerning underground operations which characterized Spanish American law.

The discovery of small amounts of gold in the stream gravels of São Paulo in 1596 led to the formation of the first mining code early in the seventeenth century.[26] Portugal at the time was affiliated with the Spanish Empire through its "acquisition" by Philip II (Philip I of Portugal) in 1580, and the new mining law, like the basic general law code governing all Brazilian affairs (called the *Leis Filippinas*), was signed by Philip III (Philip II of Portugal). Issued on August 15, 1603, the law provided that all subjects of the King could explore for mines with full latitude, provided one-fifth of the metals discovered were reserved for the Sovereign. The law's sixty-one chapters covered virtually all the situations which would arise in both the exploitation of the workings and the fiscal administration of the industry. Prospectors were to register with the local *provedor,* who would give permission for exploration upon receiving a promise that the Royal Fifth would be paid. He would then notify all interested officials. In order to validate his discovery and be permitted to work it, the prospector had to declare his claim within thirty days and have a notary *(escrivão)* register it along with the demarcation points. Long sections of the law dealt with the intricacies of measurements and the rights of codiscoverers. A miner could have only one claim within any given area approximately 4 1/2 miles square, and to keep the title, the mine had to be worked and could not be left idle at any one time for more than fifty days. Miners were encouraged to carry out public works, and they were guaranteed water and easement rights as well as the right to cut timber as needed. There were to be regular inspections of the workings, and government officials were not permitted to be partners in mining enterprises. Owners of mines could not be jailed nor their tools or slaves attached for civil suits; the judges were admonished to work out payments for debts so that the mine work would not be interrupted. In addition, anyone who interfered with mine work would be fined. Smelters and refineries were to be built by the government, which would see that they were properly supplied. All gold and silver were to be delivered to the royal casa de fundicão for refining.

These laws were carefully written, but they were designed for fairly compact districts that could be easily supervised and patrolled. In the case of Brazil, the area covered was about equal to France and Germany combined and would have needed ten provedores in order to have kept the Crown well informed. Moreover, since the São Paulo gold fields did not develop, these laws were never put completely into effect, despite the creation of a Superintendency of Mines for Southern Brazil. The second mining code, drawn up in August 1618, was a modification of the laws of 1603, and in it the duties of the provedor were increased. He was now ordered to make weekly investigations to determine whether anyone was extracting ore without paying the Royal Fifth and registering the gold dust and nuggets. The size of the first holding which a prospector could work was doubled (from 88 meters by 44 meters to 176 meters by 88 meters) though all second holdings were slightly reduced. A prize of 20 cruzados was also granted anyone who discovered mining sites. No prospector could own more than three mines, and the right to search was granted not only to Portuguese but also to Indians and even foreigners, providing they had permission to reside in Brazil. Although registered some years afterwards, the 1618 code in turn was never really implemented, and the provedores did little to encourage the mining industry. In 1694 a decree permitted the granting of titles of nobility and the awarding of decorations to men discovering mines. The finds of the great gold deposits of Minas Gerais the next year, however, could not be attributed to the offer of titles.[27]

At first the Portuguese government could hardly believe the size and significance of the gold discoveries of 1695, but when it became clear that local officials could not cope with the situation, the Governor of Rio de Janeiro, Artur de Sá, issued a mining code for the workings in Minas Gerais. This code, set forth in March 1700, was accepted by the Crown—with minor modifications and additions—and promulgated as "Regulations for the Superintendents, Crown Representatives *[guardas-môres]* and Other Crown Officials Appointed to Look After the Gold Mines" on April 19, 1702, and Governor de Sá was empowered to put it into force.[28] It remained in effect, with certain changes, for the rest of Brazil's colonial period.

The new code changed the administration of the industry by creating an extra official, the *Superintendente,* aided by a *guarda-mór* and several *guardas-menores.* Significantly, the first article ordered the Superintendent to find men causing disturbances at the mines and to punish them properly. The third article set down rules for rectifying boundary controversies, and the fourth was on how to protect the poor against rich and powerful claim jumpers. The Superintendent was to appoint a guarda-mór for rich workings and several guardas-menores to assist in the more remote places. These officials were to be well paid. They were to take a census of their districts, record their histories, and acquaint themselves with prevailing labor conditions. To ensure production they were also empowered to regulate the division of slaves for sale by the size of the workings. In order to safeguard royal revenues, they were given police and judicial rights to stop contraband and collect the Royal Fifth. In addition to being empowered to take action against tax dodgers on the basis of secret denunciations, they had authority to control all trade that might be used as a subterfuge to avoid taxation. The methods by which the collection of taxes was to be accounted for were most complex and spelled out in detail.

Discoverers of mines were given two days to report their gold strikes to the authorities, who would then demarcate the entire gold field and individual claim areas. The original discoverer could then choose the area he wanted for his own; a section reserved for the Crown was next sold at public auction for the royal account, and lots were drawn for the remaining claims by the other men present at the ceremony of demarcation. The size of these claims could be varied under several formulas; in particular they could be enlarged in proportion to the number of slaves a miner might own. Work had to be started within forty days, and dealing in mine properties was restricted.

A weakness in the concept of the law was the centralization of a large number of responsibilities in a small number of officials. In addition to acting as the regulatory agents of the industry, collecting the Royal Fifth, and overseeing the minting of new gold, they also had to encourage the technological improvement of the industry. However, officials were chosen for their administrative talents rather than for their technical abilities, partly because the scant knowledge of mine engineering in Portugal made men with such training virtually unobtainable.

The officials were not completely successful in their tasks of collecting the Royal Fifth and directing gold dust into the royal refineries and mints. Aside from the inevitable corruption, the tremendous wild area to be patrolled and the great number of tough, willful men to be policed made the task extremely difficult. It was recognized from the beginning that the collection of the Fifth would not be feasible in many instances because of the lack of facilities to refine and weigh the gold before miners spent it in the form of gold dust. (In Spanish America, in contrast, the size of most operations and the type of ore processing at the mines and patios made it easier to keep track of a substantial part of the production.) At first, all persons leaving the area had to register the amount of gold they were carrying and designate the refinery to which they were bound where the Fifth would be collected. Under this system only thirty-six persons paid the Fifth in 1701, one in 1702, and eleven in 1703! In the early years not only were the refining houses in out-of-the-way places, but so many routes, legitimate and surreptitious, led to the coast that they could not all be adequately checked. Untaxed gold in Bahia in the opening years of the eighteenth century was estimated in the hundredweights. At best, one-third of the gold actually mined was declared, and efforts to channel the precious metal through the authorized refineries and the mint at Rio de Janeiro were only partially successful. In 1710 the Governor of Rio, Antônio de Albuquerque, set a tax on the bateias used to wash gold. This was in effect a head tax on the slaves employed, and it ignored differences among mines in both yield and costs. As a result, three years later the new Governor, Dom Braz Balthazar da Silveira, reached an agreement with the various town councils to pay the Royal Exchequer a flat sum to be raised by taxes that they were to collect. The Crown balked at this short-cut, but was forced to accept it under threat of a local revolt. The next Governor, Dom Pedro de Almeida, Count of Assumar, made a compact by which the amount contributed by the town councils was to be reduced while the Crown taxed all traffic entering and leaving the gold fields. Assumar finally announced a new scheme: gold dust could circulate in Minas Gerais, but in order to leave the region it would have to be refined and taxed at one of the three refineries to be established there. Since even this scheme would entail journeys to the refineries in addition to interminable delays and even illegal exactions, the Mineiros resisted, and in 1720 a revolt broke out in Villa Rica.[29] Assumar crushed the uprising with great firmness, particularly since some of the leaders had spoken of establishing a republic; and the populace was cowed. Still, he held the refinery idea in abeyance and retained the system of tax settlements with the towns, plus various transit duties.

That same year Minas Gerais was separated from São Paulo and established as an independent Captaincy, and the new Governor, Lourenço de Almeida, was commanded to set up the refineries and a mint when he succeeded Assumar. The town councils immediately offered to raise the contribution they paid in lieu of the Fifth, but despite Almeida's recommendation to accept the offer, Lisbon held to its decision. Dom Lourenço was a brilliant negotiator with a reputation for conciliatory measures, however, and finally in 1725 he was somehow able to persuade and cajole the town council of Villa Rica to accept a smelting house and a mint which had been promised by the local authorities after the uprising of 1720 but never built.

Although Almeida in 1730 lowered the royalty rate from 20 to 12 percent, he nevertheless managed to increase the total revenues collected, with the result that few fleets had ever arrived in Lisbon richer than that of 1731. The Crown was not content, however, and ordered the old rate restored. For a while it appeared that both sides could be satisfied: Crown revenues increased and smelting houses were established at both Sabará and São João del Rei in July 1734.[30]

In 1733, however, the Crown decided to substitute a head tax for the Fifth and imposed the plan despite offers from the town councils to make large contributions each year instead. All slaves over twelve years of age, male and female, were taxed in addition to free blacks, mulattoes, and mixed bloods, who paid a tax on themselves, as did craftsmen, workingmen, and all businessmen. The refineries were closed and the inhabitants of Minas were forbidden to use coined money—only gold dust could circulate freely. The tax fell heavily on the poor and unsuccessful, for all were taxed equally whether a mine or a business were a bonanza or a marginal operation, and some mulattoes and colored women were actually forced into prostitution to raise the needed gold. Since the entire region worked on long-term credit—four to five years was not uncommon —a contraction of credit was forced, and debtors either had to borrow for a short term at higher rates or had to face bankruptcy. The adverse effects of the tax were so great that in 1750 it was repealed, the refineries reopened, and an increased town contribution accepted instead.

Reversion to collection of the Fifth at the refinery did not guarantee the Crown its due share of the mines' produce: miners were able to smuggle their gold to the coast with little risk or trouble. Perhaps less than half of what was mined was taxed. Miners justified their actions by citing bureaucratic delays at the refineries, the venality of officials, and the fact that in addition to the 20 percent tax plus 5 percent for seignorage, the Crown paid for the gold at from 10 to 20 percent less than the open market price. The height of defiance was reached in the 1730s when large-scale smuggling syndicates were formed as business companies and counterfeiting rings used former employees of the mines and refineries to establish connections in the Royal Council of Lisbon. These were broken up, but until the decline of the gold fields in the late eighteenth century, the Crown agents and a few smugglers engaged in elaborate games of cops-and-robbers with imprisonment or even death as the forfeit. Criticism of Portugal's mining laws was frequent throughout the colonial period, and yet they functioned despite the protests. In defense of the Crown it must be recognized that it was difficult to devise a code that would protect the interests of the inhabitants, control contraband activities and provide revenue for the *Real Erario.* Despite the defects in the law, villages and towns sprang up and fazendas dotted the landscape. The surviving buildings of the baroque era are clear evidence of the prosperity and taste of the inhabitants.

Since an Intendant was appointed to the Diamond District in 1734 and the region was virtually sealed off from the rest of the Captaincy, the legal situation was somewhat different in regard to precious stones.[31] Furthermore, by adopting a contract system in 1736 and by having mounted dragoons to police the area and inspectors to watch the men sifting the troughs, the Intendant made it more difficult to smuggle diamonds than gold. The proverbial royal paternalism of the Braganzas was stricter in the Sêrro do Frio region than anywhere else in Portuguese America. Again the administrators were dedicated and honest, though severe, and the disposal of diamonds in Europe was controlled by the Crown authorities in Lisbon so that, by colonial standards, the whole operation was efficient.

Among the legal provisions it is interesting to note that the contractors were not to employ more than 600 slaves in mining (overproduction was a distinct danger) and that a yearly head tax of 230 milreis had to be paid on each. Furthermore, diamonds could be sought only along certain streams and rivers; slaves caught working elsewhere were confiscated. Continuous supervision characterized the operation, and as soon as diamonds were found, they were placed in the hands of the Intendant who had specially made strongboxes for guarding them. Indeed, the Intendant was all-powerful and could expel anyone suspected of mining, buying, or selling diamonds illegally. It was even against the law to own mining instruments.

In 1753, the Crown issued a new decree by which the contractors were to be supervised directly by the royal government. Mining and trade in diamonds were even more strictly controlled; application for entrance into the Diamond District was subject to a rigorous examination; inhabitants of the district and businessmen had to be of impeccable character or they could be expelled; punishment for any infringements were draconian and informers were highly rewarded (a slave who provided information, for example, would be freed and given sufficient money to start a new life). The Crown's income rose, but so did that of the contractor, arousing the cupidity of the Crown. Finally, in 1771, the contract system was abolished; the washings were to be worked directly for the Crown.

The Diamond Code promulgated on August 2, 1771, was to go into effect on the first of the following year.[32] It centralized all diamond sales in the Royal Treasury, setting up in Lisbon an office —the Conta da Minha—headed by the Marquis of Pombal and three directors who in turn appointed three *Caixas-Administradores* (Administrators of the Strongbox) to carry out their orders in supervising all operations in the Diamond District. Their first job was to correct all the harmful and intolerable abuses practiced in the district, in particular to suppress disorders, reduce the number of slaves, and control excesses in order to protect the Crown's interests should conditions deteriorate. Political power was in the hands of an Intendant.

In many ways the code repeated the law of 1753, but it was even more stringent in its enforcement provisions. All inhabitants in the demarcated lands had to register within fifteen days of the publication of the code and also had to inventory their property and possessions; failure to do so could result in banishment to Angola in Africa for six years. All property titles were to be reregistered and subject to reapproval. These registers were to be used to control smuggling. Persons could be expelled, without the right of appeal, from the district if the authorities believed that they were not needed. Those expelled had to stay out of the district, for to return was to incur automatically the sentence of ten years in Angola. No one from the outside could enter permanently or on business without the permission of the Intendant, which required filing an application accompanied by an affidavit from his local police or justice of the peace. The length of the stay could be arbitrarily fixed by the Intendant. Article XXVI stated, simply, that no one, no matter his state, quality, or condition, is exempt from search or surveillance. The company of dragoons and the foot soldiers stationed in the region were given complete authority to do whatever they thought necessary to suppress smuggling; and no court could enjoin them from harassing a suspect. Informers were to be paid liberal rewards punctually. Suspected smugglers could be convicted on "well founded indications" and verbal secret testimony. Prosecution was to be swift and sentences summarily given. People convicted of unauthorized diamond buying had no right of appeal.

A registry of slaves was to be maintained. The administrators were to supervise the distribution of new slaves, and those slaveholders most deserving, i.e., helpful to the authorities, were to be favored. All slaves not absolutely necessary were to be sent out of the district. In addition, slaves too young or too old to work were to be sent out as well as the excess blacks during the slack seasons. Negroes were also forbidden to buy or sell anything in shops in order to reduce their opportunities to sell illicit gems.

Having secured the regulation of the lives of the people of the district, the administrators were to inspect the diamond workings. A large area was closed to diamond washings in order to simplify the policing problem; illegal miners and trespassers were to be severely punished. Rules were to be promulgated to assure the maximum productivity in both the dry and the rainy season, and a production quota was suggested based upon the former contractor's output. All slaves taken over from the previous contractor were to be registered, and particular care was to be taken with those slaves engaged in diamond washings. All slaves suspected of illegal activities were to be sold out of the district. Only buyers approved by the Crown were to buy diamonds; neither miners nor administrators could buy diamonds for resale to official buyers or the Crown. All diamonds were to be collected at the Casa da Administração and guarded day and night by troops. Elaborate precautions were spelled out for transporting the gems in a strongbox under armed guard to Rio de Janeiro for shipment to Portugal.

Despite the loopholes which developed in its system for collecting its share of the riches of Brazil, the Portuguese Crown received millions of milreis worth of gold and diamonds in the eighteenth century, and the Monarchy could boast that it neither "feared nor owed." When the calamitous earthquake of 1755 destroyed half of Lisbon, Brazilian mines helped rebuild it. The remarkable large square, the Terreiro do Paço, the network of streets around it, the Palace of Queluz, and the Monastery and Palace at Mafra—built by João V to rival the Escorial—all testify to the enterprise and climate of wealth prevailing in the country. The sumptuous royal patronage of arts and letters (one remembers the decade when Domenico Scarlatti was Court composer in Lisbon) would not have been possible without the riches of Minas Gerais.

As the new century began, the Lisbon government, and particularly D. Rodrigo de Sousa Coutinho, came to feel that a new mining code should be drawn up for Brazil. The Minister and his mineralogical adviser Manoel Ferreira da Câmara Bethencourt e Sá[33] discussed the question frequently and the latter was asked to draft a document to succeed that of Pombal. Since Câmara was a Mineiro who had visited the mines of Europe, it was natural that he was selected for the task. His conclusions, with some modifications, were incorporated in the *Alvará* (Court Order) of May 13, 1803, which made nine important points:

1. Do away with the circulation of gold dust in all the Captaincies where it has been used instead of currency.
2. Transfer the Casa da Moeda (Mint) of Rio de Janeiro to the Captaincy of Minas Gerais and that of Bahia to the Captaincy of Goiás.
3. Establish *casas de permuta* (smelting houses) for the exchange of gold and purchase of diamonds in all the mining districts.
4. Reduce the Royal Fifth to a Tenth for the benefit of the King's vassals.
5. Borrow a million and a half cruzados in order to mint silver and copper coins (as well as gold).
6. Divide up unclaimed sites of the Diamond District so that they could be worked and taken advantage of.
7. Create a Junta called the Administrative Junta of

Mining and Minting of the Captaincy of Minas Gerais.

8. Name a General Intendant of Mines to put into execution everything set forth in the alvará as well as other wise and appropriate measures.
9. Establish a school of mines in the Captaincy.

The Prince Regent, Dom João, warmly approved of the alvará and instructed the Governor (who was also the Captain-General) of Minas Gerais to assume the presidency of the Administrative Junta, with the new Intendant as Vice-President and the Ouvidor Geral, or Senior Royal Judge, as Legal Counselor. The other members were to be the Provedor, or Director of the Mint, two Deputies versed in mineralogy, one or two mining engineers named directly by the Crown, and two intelligent and well-established local miners. The two Deputies should be elected every three years by the Junta and the office given such prestige as to make it eagerly sought after. Dom João was particularly enthusiastic about a school of mineralogy and metallurgy similar to those in Freiberg and Chemnitz which had been of such help to the kingdom of Saxony.

The departure of Dom Rodrigo from the Ministry in 1803 and the pressure brought by the Diamond District officers to have the old *Real Extração* bureau in Tijuco retained—as well as Napoleon's bullying tactics with the Braganzas—delayed implementation of any change until December 1806 when Câmara finally became de facto Intendant. On November 2, 1807, together with the Governor, the *Escrivão Deputado* (Deputy Registrar), and the *Ouvidor da Comarca* (District Judge), Câmara signed a *parecer,* or opinion, regarding the Alvará of 1803 which he had largely prepared four years previously.

The situation was a delicate one for a man coming into a region that was jealous of its prerogatives and suspicious of change. An Administrative Junta of Mining and Minting was formed, and the three principal members, with Câmara in accord, found that the Captaincy did not have sufficient funds to obtain a loan to buy a substantial amount of silver and copper for new coinage and recommended rather modest substitute measures instead. Gold dust did cease to circulate, however. Assurances were given to the local inhabitants owning claims and no unknown people were allowed in to seek diamonds. Câmara was very much in charge, however, and he managed matters with tact and common sense. The travelers Saint-Hilaire and Mawe were impressed by his paternal approach to problems and his basic concern for improving methods of production.

Notes

1. For collections of Spanish law codes, see Marcelo Martínez Alcubilla, ed., *Códigos antiguos de España: Colección completa* (Madrid, 1885), and *Códigos españoles* (Madrid, 1872). Two of the early codes were translated into English by Samuel P. Scott in *The Visigothic Code (Fuero Juzgo)* (Boston, 1910) and *Las Siete Partidas* (Chicago: Comparative Law Bureau of the American Bar Association, 1931). Also see John A. Rockwell, *A Compilation of Spanish and Mexican Law in Relation to Mines and Titles to Real Estate* (New York, 1851), and Henry W. Halleck, *A Collection of the Mining Laws of Spain and Mexico* (San Francisco, 1859), as well as the interesting summary entitled "Institutes of the Civil Law of Spain," in Joseph W. White, *A New Collection of Laws, Charters and Local Ordinances of the Governments of Great Britain, France, and Spain, Relative to the Concessions of Land in their Respective Colonies,* vol. I (Philadelphia, 1839), pp. 352–356.

2. The term *civil law* applies to the Institutes of Justinian which were appealed to in cases not covered by Spanish enactments. Since the *Fuero Juzgo* (the Visigothic Code) was silent on the matter of mining, civil law prevailed in medieval Castile.

3. Francisco Xavier de Gamboa, *Commentaries on the Mining Ordinances,* chap. 2, par. 1; chap. 4, pars. 2 and 3, reprinted in Rockwell, op. cit., pp. 124 and 163; *Code of Justinian,* book XI, title vi, laws 1 and 3 [in Samuel P. Scott, trans., *The Civil Law* (Cincinnati, 1932), vol. XV, p. 169].

4. *Las Siete Partidas,* law 5, title xv, part II, and law 11, title xxviii, part III; also reprinted in Rockwell, op. cit., 126 n.

5. Originally in the *Ordenamiento de Alcalá,* it was copied into the *Recopilación de Castilla,* law 3, title xiii, book VI, and in the *Novisima Recopilación,* law 1, title xviii, book IV. (Also in Rockwell, op. cit., p. 112.) Also see Gamboa, op. cit., chap. 2, par. 7, reprinted in Rockwell, op. cit., p. 126.

6. Gamboa, op. cit., chap. 2, par. 8. Also see Rockwell, op. cit., pp. 112–113 and 126, and Halleck, op. cit., p. 5.

7. José María Ots Capdequí, *Estudios de historia del derecho español en las Indias* (Bogotá, 1940), pp. 15–57. Also see Juan de Solórzano y Pereyra, *Política indiana* (1647), book VI, chap. 1 (reprinted in Madrid and Buenos Aires by the Compañía Ibero-Americana de Publicaciones, 1930). While some mines—such as copper mines in Cuba—were temporarily reserved for the Crown, others were

permanently incorporated into the Royal Patrimony, such as the mercury mines of Huancavelica, Peru, which provided the mercury for the silver amalgamation process at Potosí.

8. This and the following information is from Arthur S. Aiton, "Ordenanças hechas por el Sr. Visorrey Don Antonio de Mendoça," *Revista de Historia de América,* no. 14 (June 1942), pp. 73–80.

9. *Ordenanzas del Perú,* reprinted in Emilio Tagle Rodriguez, *Legislación de minas* (Santiago de Chile: Imprenta Chile, 1918). Gaspar de Escalona made a compendium of these ordinances with marginal comments, and Tomás de Ballesteros made a complete compilation. Ordinances were also written by Juan Matienzo, a judge of the Audiencia of La Plata. As early as 1562, Licentiate Polo drew up some ordinances for the mines of Huamanga.

10. Gamboa, op. cit., chap. 1, par. 7. Also see Rockwell, op. cit., pp. 119–120.

11. *Recopilación de las leyes de las Indias,* law 3, title i, book II.

12. Aiton, op. cit., pp. 79–80.

13. Ibid., p. 79 n.

14. Halleck, op. cit., pp. 6–15.

15. Ibid., pp. 17–24.

16. For a discussion of this problem, see Gamboa, op. cit., chap. i, pars. 1–4, in Rockwell, op. cit., pp. 119–120.

17. Halleck, op. cit., pp. 69–81.

18. See Gamboa's discussion, op. cit., chap. i, pars. 1–6, and Tagle Rodriguez, op. cit., p. 44.

19. For an excellent survey of the background of the ordinances and an analysis of their contents and effects, see Clarence H. Haring, *Spanish Empire in America* (New York: Oxford University Press, 1947), pp. 263–267.

20. Halleck, op. cit., pp. 189–308.

21. Ibid., pp. 306–12.

22. In his *Guía histórica,* Pedro Vicente Cañete writes a detailed history of the Mint of Potosí, including the construction, costs, machinery, work that was carried out, directors, workers, and accomplishments.

23. Diffie, *Latin American Civilization: Colonial Period* (Harrisburg, Pa.: Stackpole Sons, 1945), p. 369.

Don Fausto de Elhuyar y Zubice, the discoverer of tungsten, was appointed in 1786 to serve as Director General of Mining in New Spain and to head the Royal College of Mining in Mexico City. Humboldt called him "a famous savant in the annals of the chemical sciences." (*Consejo de Minería, Madrid.*)

Don Andrés Manuel del Río, the discoverer of vanadium, was one of the outstanding professors at Mexico's College of Mining. His treatise on mineralogy was one of the most notable of its time in any language. (*Palacio de Minería de México.*)

José Bonifácio de Andrada e Silva, a trained mining engineer, was known as a geologist throughout Europe since he was the discoverer of several previously undescribed minerals. Returning to his native Brazil in 1819, he became the guiding spirit of the movement for independence in 1822. Portrait from S. A. Sisson, *Galeria dos Brasileiros Illustres* (1861). (*The Hispanic Society of America.*)

24. See *Los Bancos de rescate de plata* by Pilar Mariscal Romero (Madrid, 1964), published by the Bank of Spain. This study which refers particularly to the Viceroyalty of New Spain is the first one dedicated to this subject.

25. See Cañete, op. cit., chap. VII: "It was the Governor of Potosí, Don Jorge Escobedo y Alarcón, who carried out the transformation of the private Redemption Bank into the Royal Bank of San Carlos, managed by the State. With this new entity in operation, all sorts of difficulties in silver mining could be avoided. Escobedo personally wrote and put into practice a detailed and adequate set of regulations for the Bank, which clearly demonstrated his administrative gifts. This distinguished civil servant also reorganized the administration of the Mint and other important offices of the Crown in the far-flung jurisdication of Potosí. For his high qualifications and excellent service in the Viceroyalty of Peru, King Charles III recalled him to Madrid with the rank of Minister of the Council of the Indies."

26. For a copy of this code and other laws affecting mining, see *Collectanea de scientistas extrangeiros* in the *Publicações do centenário em Minas Gerais,* pp. 213–228. Also see W. L. Eschwege, *Pluto Brasiliensis* (Berlin: G. Reimer, 1833), and João Pandía Calógeras, *As minas do Brasil e sua legislação,* 3 vols. (Rio de Janeiro, 1904–1905), for further comments. There is interesting background material in Damião Peres, *Antecedentes da legislacão concernente ao ouro do Brasil nos séculos XVI e XVIII* (Lisbon: Academia Portuguêsa de História, 1956. Estudos de hístoria luso-brasileira).

27. For mining activities in the state which supplied most of the prospectors see Francisco de Assis Carvalho Franco, *História das minas de São Paulo* (São Paulo, 1964), and Alfredo Ellis, *O ouro e a Paulistânia,* (São Paulo, 1953). A summary of the early years in Minas is found in Charles Boxer, *The Golden Age of Brazil 1695–1750* (Berkeley: University of California Press, 1962), pp. 30–37. For fuller details consult the studies of the American geologist Orville A. Derby who spent many years in Brazil: "Os primeiros descobrimentos de ouro em Minas Gerais," *Revista do Instituto Histórico e Geográfico de São Paulo,* vol. 5 (1899–1900), pp. 240–278, and Orville A. Derby "Os primeiros descobrimentos de ouro nos distritos de Sabará e Caeté," ibid., pp. 279–295. One of the most recent studies is Augusto de Lima's *As primeiras vilas do ouro* (Belo Horizonte, 1962).

28. *Collectânea,* pp. 231–243. Also see the comments of Boxer, op. cit., chaps. 2 and 7, and Caio Prado, Jr., *Colonial Background of Modern Brazil,* pp. 195–207.

29. For a discussion of this uprising and the role of Filipe dos Santos see the detailed two-volume study of Diego de Vasconcelos, *História antiga de Minas Gerais (1703–1720)* (Rio de Janeiro, 1948), and Téofilo Feu de Carvalho, *Felipe dos Santos Freire na sedição de Vila Rica: Ementário da história de Minas* (Belo Horizonte, 1933).

30. Boxer, op. cit., p. 197.

31. For the laws regarding the Diamond District consult Joaquim Felicio dos Santos, *Memórias do distrito Diamantino* (Rio de Janeiro, 1956). (The first edition of this classic work appeared in 1868.) Also see Augusto de Lima, Jr., *História dos diamantes nas Minas Gerais* (Lisbon and Rio de Janeiro, 1945), and João Camillo de Oliveira Torres, *História de Minas Gerais,* 5 vols. (Belo Horizonte, 1961), especially vols. I–III, pp. 1–798.

32. Reprinted in João Camillo de Oliveira Torres, *História de Minas Gerais,* pp. 266–290.

33. For the role of Manoel Ferreira da Câmara Bethencourt e Sá who became Intendant in 1806, see Marcos Carneiro de Mendonça, *O Intendente Câmara,* 2d ed. (Rio de Janeiro, 1933; São Paulo, 1958).

X

The Permanent Settlement of Iberians in America

Throughout the three centuries of Iberian dominion in the Indies, the majority of the settlers who went there characteristically did so in a spirit of permanence, intending to put down firm roots in the lands that became their new homes. This was true of those who emigrated voluntarily to improve their fortunes as well as of public functionaries—except, of course, for the few high officials who wielded royal and ecclesiastical power.

Soon after men and women from the Iberian Peninsula arrived in the New World, a feeling of being American was born in them, linking them to their new homeland. These settlers were called *Indianos* in Spain, and as such, they wished to live and die in their new land. An exemplary case is that of Hernán Cortés. When the conquistador died in Castilleja de la Cuesta, in the Province of Seville, he left instructions in the first clause of his will that he be buried "in my City of Coyoacán." Salvador de Madariaga refers to Cortés in his biography as the "conquered conqueror" and claims that Cortés was the first man who felt Mexican patriotism and was possessed of a vision of Spaniards and Mexicans living peacefully together. To that end Cortés founded and endowed hospitals, religious institutions, and centers of learning to provide intellectual and spiritual leadership for the natives of New Spain.

This phenomenon of permanent settlement in the American domains produced situations quite unlike those of "colonies" in the usual sense of the word. It also highlights some very characteristic aspects of life in the Spanish Viceroyalties and Portuguese Captaincies of the New World. In the first place, it explains the fact that the actions of individuals and also of the authorities who represented the Crown were bent toward creating lasting institutions and life-styles similar to those of the mother country, although adapted to the special circumstances of the new lands. This tendency to create a permanent, stable life demonstrates that most of what happened in the New World after the arrival of the first transplanted generation was the work of *Americans,* carried out by them with the nature and temperament imposed by the mingling of races and the conditions of American life. These new citizens of America were the men who, with full responsibility, created and set the tone for institutions; who organized and benefited from agricultural, mining, and commercial exploitation; and who were responsible for dealing with the natives. They also built churches, educational centers, and hospitals, and organized expeditions both inside and outside their own territories. One result of this activity was the opening of sea routes from Acapulco to the Philippines and the Moluccas and their maintenance for almost 300 years.

Hence neither the Spaniards nor the Portuguese of today should boast overmuch of their great and often glorious accomplishments. By the same token, neither should their counterparts in the Americas excuse themselves by blaming the home governments for the things their ancestors did that may now offend their sense of justice. These terrible things happened when the Laws of the Indies and other regulations intended to protect the natives were either violated or wrongly applied. Also, at times the institutions created for the particular circumstances of the New World were perverted, such as the practices of encomienda and mita and the trade in Negroes and mulattoes contrary to the Crown's express regulations.[1] On the other hand, this sense of living not in a transitory settlement but in one's own permanent home, of working for oneself and one's descendants, explains the grand scale on which the settlers built their cities and the care they lavished on them, especially the capitals which were the centers of the Viceroyalties and Captaincies.[2]

The number and size of the cities the Spaniards founded both on Caribbean islands and on the American mainland, starting immediately after the Conquest, is impressive. They provide clear evidence of the Spaniards' attitude toward Columbus's great discovery for the Crown of Castille and their intention of transplanting their lives and their culture to the New World.

The first city, founded by Columbus himself, was Santo Domingo on the island of Hispaniola. For its role as the gateway to America, the Spaniards established there the first legal and religious authorities over the newly discovered lands. And it was from Santo Domingo that new expeditions were organized to conquer the islands and the lands surrounding the Gulf of Mexico and the Caribbean Sea. Cities were founded on other islands as the islands were discovered and conquered. Havana, for example, replaced Santo Domingo as the principal port and became the base for expeditions to make new discoveries. The same process continued throughout the continent as more and more lands were discovered.

After the Conquest of New Spain, the organizing spirit of Hernán Cortés—a product and a representative of the Renaissance in Europe—was responsible for carrying out in an astonishingly short time the exploration of the country and its coasts, the establishment of cities, and the building of ports. That is why Cortés founded the city of Veracruz on the coast at the very moment he landed and established there the first municipal government on the mainland of America. Very shortly thereafter, Campeche, on the Gulf of Mexico, and Tehuantepec, Acapulco, Zihuatanejo, Manzanillo, and Barra de Navidad, on the Pacific, were operating as ports where ships were built and from which new expeditions set sail. In 1527, six years after the Conquest of Mexico, an expedition of three ships sailed from Zihuatanejo under the command of Alvaro de Saavedra and crossed the Pacific Ocean to the Molucca Islands.

The founding of these ports in New Spain was followed by the building of others on the Caribbean shores, such as Portobello in Panama and Cartagena de Indias in modern Colombia. In all these cities an important type of military architecture developed, marked by new designs in fortifications, walls, citadels, and arsenals, to be used as defense against attacks by pirates and buccaneers who were attracted by the ports' wealth of precious metals. The defensive works of the fortress of San Juan de Ulua at Veracruz were notable; Hawkins and Drake beat against them in vain in 1567. Equally strong were the fortifications at Cartagena de Indias, the target of several daring and sustained attacks.

The Church effected the most important architectural movement, however, filling the New World with great cathedrals and beautiful churches and convents. The first cathedral was begun in Santo Domingo in 1512 and finished in 1541. It is probable that

Columbus is buried there, although Seville still disputes this claim. Other cathedrals and religious edifices, some of extraordinary beauty, were scattered over the Caribbean Islands and the mainland. In this work an important role was played by the three mendicant orders, the Franciscans, the Dominicans, and the Augustinians who arrived immediately following the Conquest and later were joined by Jesuit fathers.[3]

The building of churches and their related structures depended of course upon the economic well-being of the Viceroyalties, and as Sylvester Baxter pointed out seventy years ago: "Almost universally throughout Mexico the wealth of the Church and the consequent splendor of the ecclesiastical architecture were based upon the marvellous mineral production of the country."[4] This is not the place to review colonial architecture as a whole, but a few words should be said about some of the religious buildings erected in or near the areas where silver or gold were actually mined. Some of them were "so manifestly costly and so elaborate in their adornment as to seem a sort of magical and spontaneous growth from the treasures in the ground below."[5] While miners were anxious to make fortunes, they never forgot that God was watching over their destinies, and, consequently, they contributed lavishly to the erection of churches, chapels, and shrines. Guanajuato, a region remarkable for its silver deposits and one of the world's great mining centers, possesses a number of buildings of unusual beauty such as the church of San Francisco with its eighteenth-century sacristy and baptistry and carved confessionals, and the huge temple of la Compañía, erected between 1747 and 1765. A blending of baroque and Churrigueresque, it was supervised by the Jesuit Father José Joaquin de Sardaneta y Legazpi, a relative of the wealthy mine owner, the Marques de Rayas. The elaborate baroque style is seen to wonderful advantage in the facade of the church of San Diego, erected in the 1780s, but the most remarkable building of the region is the church of San Cayetano built in the 1770s and 1780s a few miles from the capital city. The building, resting on top of the renowned mine La Valenciana was built through the zeal of the owner, the Conde de Rul. La Valenciana was the richest of the Guanajuato mines and is said to have yielded up to 800 million dollars worth of silver. Of course its site is dramatic, but the unknown eighteenth-century architect was equal to the task of designing "a temple of riches built to proclaim the glory of God with the might of man." The mudéjar exterior and the interior masonry are worked with consummate skill. Each stone cutter was apparently given free rein in the design of the ornament of the piers and arches and thus great variety is obtained. The pulpit is a magnificent example of inlaid work in the Querétaro style. Everyone engaged in the working of La Valenciana took pride in San Cayetano, and in its heyday each employee contributed a weekly piece of ore or *piedra de mano* for the upkeep and services of the church, which is said to have amounted to $50,000 annually.[6]

Perhaps the outstanding mining church in Mexico is San Sebastián y Santa Prisca at Tasco in the state of Guerrero, already mentioned in this study. This extraordinary masterpiece, completed in 1757, was erected through the efforts of D. José de la Borda, whose fortune derived mainly from mining properties in Tasco. No expense was spared in the decoration and furnishing of this rococo church set among giant precipices 5,000 feet above the sea. The facade and towers of fine-grained brown stone are elaborately embellished with sculpture, and the interior contains a dozen memorable chapels with splendid altar retables of gilded carved wood and polychrome sculpture. There are also magnificent mural decorations throughout the church by the gifted Mexican painter Miguel Cabrera. The overall concept and the creamy carved stone against a dark salmon background are most harmonious. Santa Prisca's sacristy is equally lovely and contains fourteen panels by Cabrera of Christ and the Virgin. As one gazes at the deep blue, orange, green, and white tiles of the dome, which has been compared to a tropical flower, the inscription on the frieze of the drum—"Gloria á Diós en las alturas" (Glory to God on the heights)—seems the only possible summation of its beauty.

One of the most notable gardens of eighteenth-century Mexico, el Jardín de la Borda at Cuernavaca, was designed for the son of the mining Maecenas, D. Manuel de Borda, and its hillside terraces, vistas, pavilions, fountains, and basins reflect the aesthetic sensibilities and taste of the colonial mining gentry.[7]

Several other mining regions deserve mention, such as Zacatecas in the high plateau of north central Mexico. The principal settlement was a remote town in a ravine 8,000 feet above sea level built over a rich vein of silver discovered by Juan de Tolosa in 1546 and granted the title "Very Noble and Loyal City" in 1585. In 1803 Humboldt said that Mexico's most productive mines were to be found there. The rich brown or red stone of the area has a fine-grained texture which enabled local artisans to produce elaborate ornaments without difficulty. The results are seen to great advantage in the cathedral which was completed in the middle of the eighteenth century. The exterior is noteworthy for its intricate baroque carving, and there is a distinct "Indian" flavor to the design; indeed, the low reliefs have an almost Aztec look.[8] San Luis Potosí, capital of the state of the same name in the eastern table lands of north central Mexico, was founded in 1586. Its surroundings are celebrated for rich mines, and the Catorce district still has some of the principal silver lodes of Mexico. One of the loveliest churches is Nuestra Señora del Carmen whose facade is one of the most elaborately sculptured in New Spain. The remarkable domes in the mudéjar style, decorated with green, yellow, blue, and white tiles, are capped by two beautiful lanterns, while the intricately carved tower on the north side of the west facade is crowned with a blue and yellow pyramid. The large polychrome screen in the cruciform interior is by the gifted eighteenth-century Mexican architect, Eduardo de Tresguerras. The Antigua Real Caja, the former Royal Treasury, has one of the finest courtyards of the Viceregal period.[9]

Another center of colonial architecture associated with a mining region is the city of Oaxaca, "the noble city, par excellence," lying at an altitude of 5,000 feet in a valley flanked by ranges of the Sierra Madre del Sur. Conquered in 1552 by Juan Núñez de Mercado, one of Cortes's captains, the place is notable for its green onyx buildings and above all for the lovely cathedral dating from the sixteenth century. Fray Antonio Vázquez de Espinosa in his *Descripciones de la Nueva España en el siglo XVIII* called it "among the best and most perfect in the Indies."[10]

The many other remarkable buildings of the principal centers in New Spain resulting from the Viceroyalty's mineral wealth are far too numerous to describe, but a few words should be said about the Escuela de la Minería (the School of Mines) in Mexico City. Begun in 1797 for the purpose of training mining engineers, this much-needed school, the work of the architect Manuel Tolsá, is a handsome neoclassic building whose harmonious proportions are both elegant and impressive, and as George Kubler points out: "The staircase of converging ramps is the most grandiose colonial structure of its kind in America."[11]

Farther to the south we find the same phenomenon as in New Spain. Prosperity and cultural advancement came to the Real de Minas de San Miguel de Tegucigalpa, today the capital of Honduras, through the silver mines of the region which were discovered in 1578. However, it was not until the eighteenth century that the colonial apogee was reached there. Then master craftsmen came from Guatemala and Comayagua to remodel the large parish church (today the cathedral), and although the structure was made solely of local brick and plastered over and whitewashed, and the exterior with its deeply grooved pilasters has a simple aspect, the inside of the building takes one into the ornamental splendor of the period. As Pal Kelemen has written: "The interior preserves one of the most exquisite rococo altars in the New World—a lacy shimmering fantasy of carved wood, featuring archangels guarding a figure of the Virgin. A fragile four-wheeled cart perches atop the marching pulpit—evidently the triumphal Chariot of Faith."[12]

New Granada (present-day Colombia) was one of the chief gold-producing regions of the colonial period. There were numerous placer mines with the main centers, as mentioned above, being Antioquía, Barbacoas, and Chocó. Many of those who prospered from gold mining, however, came to live in Popayán, a city founded in 1536 on a volcanic terrace 1,000 feet above the Cauca River at an altitude of 5,500 feet. The inhabitants, predominantly of European ancestry, had a high standard of living, and the town grew to be a rich, aristocratic center of colonial trade. Of the several important churches in Popayán, the most striking is San Francisco, erected between 1775 and 1795. The facade, by the architect Antonio García, is by general consent the handsomest built in New Granada during the eighteenth century, and Kubler speaks of "its ripples in broken silhouette curves like a Borrominesque church."[13] The President of the Audiencia wrote in 1727 that the owners of the Chocó mines were all residents of Popayán; their wealth is suggested by the fact that in 1778 the Royal Fifth amounted to 18,070 castellanos.[14]

The discovery of silver in 1545 in Alto (Upper) Peru (present-day Bolivia) was one of the great finds of modern times, and for more than half a century Potosí was the most fabulous source of silver the world had ever known. In the symbolic language of the period: "Potosí is the precious heart of the Spanish Indies and its silver the life blood of the Casa de Austria." Yet, because it is situated some

13,000 feet above sea level—well above the timber line—the architecture of Potosí is severe and the town has an austere look. In a community plagued by high winds and whirling dust, the tendency to build narrow streets and thick walls is understandable, particularly in an area where earthquakes might occur. Life was hard; yet despite the setback suffered from the disastrous flood of 1626 (caused by the breaking of one of the water-power reservoirs), churches and residences were built throughout the colonial period. Among them one may single out La Merced with its facade dating from the 1680s. Related to the school of Cuzco, the *portada* (facade) with its double-storied columns seems to have been carved in wood rather than stone. The central space of the second story has an unusual vaulted niche between pyramids on pedestals, while the shields and urns in the niches between the columns are characteristic of the order. Curiously enough, the same ornamental symbols are used in the plaster work of the arches of Santo Domingo in Oaxaca, New Spain. Potosí had its own style of planiform architectural decoration which Kubler has compared to "the diagonal cuts of woodcarving,"[15] and the masterpiece of the school is San Lorenzo (1725–1744). This eighteenth-century church has a remarkable portada with elaborately carved surfaces and caryatids with a distinct Quechuan flavor crowning the twisted columns, probably the work of mestizo or Indian craftsmen. The blending of Spanish and indigenous cultures reached a high point in this Andean temple.[16] The most prominent secular bulding from the end of the colonial period is the Casa Real de la Moneda, or Royal Mint constructed between 1753 and 1773. It is a large building, relatively functional for its period, and its simple arched windows, iron balconies, traditional courtyard, and sober two-storied portada have a dignity in keeping with its use. A bow to indigenous design is noticeable in the "placas seriadas," which are the principal ornament of the doorway.[17]

Since the climate and living conditions of Potosí were inhospitable, another city arose in the vicinity to guide and develop the religious, legal, political, and educational activities of the vast region, whih embraced present-day Bolivia, Chile, Argentina, and Paraguay. Many successful miners owned houses in this town, which has gone by four names during its history: La Plata, Charcas, Chuquisaca, and Sucre. Today it is the de jure capital of Bolivia. While Potosí has an almost exaggerated Castillian appearance, La

Plata at only 8,500 feet is rather Andalusian in feeling. Here there are trees, flower gardens, and orchards, and it was natural that it should become the intellectual center of Alto Peru. The law courts (Audiencia), the Archbishop's See, a Municipal Cabildo, and "The Illustrious University of St. Francis Xavier" dating from 1624 all give La Plata a special flavor. The prestige of these entities in colonial days was so great, in fact, that the saying was: "To mention the Audiencia of La Plata, or the University of St. Francis Xavier is equivalent to stating that one has influence with the King of Spain and the Council of the Indies." The interdependence of Potosí and La Plata, surrounded by farms supplying the toilers of the higher city, is an excellent illustration of the radial influence mining had on settlement and economic growth. Those who made money by extracting ore from the earth gave generously to churches and convents in La Plata, and there are more remarkable colonial buildings there than in the Imperial City. The spacious cathedral, begun in 1571, is noted for its splendid lateral baroque portada dating from the 1690s, but the late seventeenth-century Iglesia de las Mónicas interests us perhaps more because of the Indian elements on the facade; the special longitudinal incisions accenting the chiaroscuro effect; and the decorative themes of pomegranates, shells, flowers, sirens, and even dogs in the spandrels of the entrance arch. The Church and House of the Oratory of St. Philip Neri (1795–1800) is the most important late-baroque monument of La Plata, and since Buenos Aires had become a Viceroyalty in 1776, French Bourbon influence via Spain is visible in the rococo retables, pulpit, and confessionals. The handsome exterior, however, has a neoclassic restraint which seems to presage the Age of Reason and republican ideas.[18]

Painters and sculptors of the colonial period have not been referred to, though they were legion, since they would lead us too far afield, and again, churches and public buildings had easily documented ties with the lives of miners and mine owners. However, there is one painter who, because of his continuous association with Potosí and the surrounding towns in Alto Peru, should not be passed over. This was Melchor Pérez Holguín, born at Cochabamba about 1660, an artist who was both cosmopolitan and local. One can trace the influence of Flemish painters and engravers in his subject matter and his Mannerist lighting, and the derivation of some of his coloring and draftsmanship from the Italian, Bernardo Bitti, who visited Alto Peru around 1600. At the same time Holguín was very personal in his anecdotal style and quite Andean in feeling. (The Virgin dressed as an Indian is shown washing clothes on a wooden board.) In short, no one would mistake him for a European artist. In *The Last Judgment* (1708) at San Lorenço, "a vast Bosch-like allegory" inspired by a Flemish print, Holguín reserved a small space for himself and five Indians in back of the angels in heaven. An extremely detailed canvas depicts the Viceroy Fray Diego Morcillo Rubio de Auñón entering Potosí in 1716, and this fascinating "documentary" is today in the Museo de América in Madrid. As Martin Soria says of his work:

> Holguín's style presents Mannerism coated in Rococo, a mixture original not only because of its atavism. For invention, personality, temperament, expression, and precise execution, Holguín may be placed ahead of any Colonial painter born in South America.[19]

Turning to music, it is well to remember that while the capitals of the Viceroyalties, the episcopal cities, and the seats of the Audiencias attracted the principal composers and singers of the colonial period, the mining regions challenged them to a surprising degree. From the very beginning the Mexican and Andean Indians showed an affinity for European polyphony. Oaxaca was especially known for its musicians, and in the seventeenth century a Zapotecan Indian, Juan Matías, who had been trained locally, was considered superior to the "able and illustrious competitors from both the capital city and from Puebla" who vied with him for the position of chapelmaster.[20] In 1691, one of his successors, Matheo Vallados, put music to some of Sor Juana Inés de la Cruz's *villancicos,* which were sung in the cathedral in November of that year. But the lure of Oaxaca is best illustrated by the fact that the cosmopolitan chapelmaster of the Cathedral of Mexico City, Manuel Zumaya, who in 1711 wrote the first North American opera, *La Parténope,* for the Viceroy, the Duke of Linares, was induced in 1732 to leave the capital city for the great mining town in southwestern Mexico. The libretto of *La Parténope* was published, as well as a play, *Don Rodrigo* (1708), and the versatile Zumaya was also known for cultivating "the sacred sciences." A number of magnificats and villancicos have survived, and they show him to have been a skilled musician, perhaps the outstanding composer of eighteenth-century Mexico.[21] One of his villancicos, scored for a solo voice with instruments and a figured bass, *A la Asunción de Nuestra Señora,* may be heard in a recording entitled *Salve Regina.*[22]

In the Viceroyalty of Peru, two composers must be cited: Gutierre Fernandez Hidalgo and Juan de Araujo. The former spent his early years in Bogotá, was in Quito in 1588 and 1589, and after seven or eight years at Cuzco, advanced to La Plata, the residential city of Potosí's mining elite. Here he spent the rest of his life singing, composing, teaching at the St. Elizabeth of Hungary Seminary, and officiating as chapelmaster of the cathedral. A project to publish five volumes of his compositions (masses, magnificats, hymns, Holy Week Office music, and motets) in the Netherlands unfortunately did not materialize, although he had pledged funds amounting to five years of his chapelmaster's salary for the purpose. As luck would have it, the manuscript was lost on its way to Europe, perhaps captured and thrown into the sea by unlettered English pirates. Happily, a few compositions surviving from the Bogotá period give an idea of his talent, and a beautiful magnificat (Tono IV), alternating plain song and polyphony, is also recorded in the collection *Salve Regina.*

The other musician, Juan de Araujo, who also worked in La Plata "was the peer of any Peruvian composer of the *virreinato.*"[23] Arriving in Lima as a youth from Spain, he attended the University of San Marcos and studied music in the "City of the Kings." After serving as chapelmaster in the Cathedrals of Panamá and Lima (1672–1676), he proceeded to La Plata where he remained until his death. Although these were not the palmiest days of Alto Peru, the cathedral in which Araujo directed the music was sufficiently well-off to vote funds to equip a whole fleet to subdue *el enemigo pirata inglés,* whose marauding was causing trouble along the coast. Hundreds of compositions by Araujo have survived in the archive of the cathedral and a few have been recorded. One is a curious Negro-dialect villancico, another a hymn to St. John the Baptist, *Ut Queant Laxis,* a simple, moving piece introduced by a Gregorian chant followed by a chorus with an accompaniment of stringed instruments and organ. Both compositions are included on the *Salve Regina* disc.

Just as mining in Spanish America led to the creation of churches and other works of art, the mineral wealth in Portuguese America fostered architecture, sculpture, and music, especially in Minas Gerais. Colonial Brazil produced a whole school of talented architects, sculptors, painters, and composers. Some of the most prominent architects were Antônio Francisco Lisbôa, José Fernandes Pinto de Alpoim, Antônio de Sousa Calherios, and Manuel Francisco de Araujo.[24]

Villa Rica, being the largest and wealthiest of the mining towns, had the greatest number of notable buildings. Representative examples include the Igreja de Nossa Senhora do Pilar de Ouro Prêto (1720), the elliptical mother church where the Governors took office; the small chapel which honors Padre Faria, one of the first discoverers of gold deposits; the large church in the parish of Antônio Dias, named for one of the original miners; and the church of Santa Efigênia (1727). It is interesting to recall that Santa Efigênia was built by the black Irmandade (confraternity) of the Rosary in collaboration with the black workers of the mines of Encardiceira. The Igreja de Nossa Senhora do Rosário dos Prêtos, with an unusual double oval plan, also was built by the Irmandade of the Rosary.

The outstanding religious monument of Villa Rica, however, was the Igreja da Ordem Terceira de São Francisco, dating from 1766. This magnificent structure was designed by the greatest and most versatile architect and sculptor of Brazil's colonial period, the mulatto Antônio Francisco Lisbôa, son of Manuel Francisco Lisbôa and a Negro slave. A disease, presumably leprosy, mutilated his hands in middle life and led to the sobriquet by which he is best known, *Aleijadinho,* "the Little Cripple." Partly because he was totally in charge of the building from its inception to its completion in 1794, the church is unrivaled for unity and beauty.

Aleijadinho also worked on the Franciscan church of São João del Rei, but he is best known for the celebrated Stations of the Cross and especially the Twelve Prophets which he carved for the Santuário-Basílica do Senhor Bom Jesus de Matozinhos. The more than life-size images of the prophets in soapstone, completed in 1805, are among the most remarkable sculptures in the entire hemisphere. And they are striking testimony to the lengths to which grateful entrepreneurs in the mining district would go in giving thanks to God.

Other settlements besides Ouro Prêto also contain impressive buildings. Ribeirão do Carmo, renamed Mariana in 1745, has a fine cathedral basilica (1720), a very handsome capitular house (1756), and a church of the Third Order of São Francisco with an elaborately decorated interior and a lovely painted ceiling. Its municipal hall, begun in the late eighteenth century, is one of the most beautiful secular buildings in Minas Gerais. Sabará, in addition to its mother church (1710), has a small church with a fascinating oriental flavor, possibly designed by a priest who had lived in Macao. The church, called Our Lady of "O," was so named because the intoned antiphons in its ritual of the vigil of the Nativity begin with the exclamation "O." In São João del Rei a richly decorated Carmelite temple contains an especially fine consistory with furniture of jacaranda wood, while the church of São Francisco of Assisi, erected between 1774 and 1804 and designed by Aleijadinho, has handsome sculptured soapstone on the exterior.

Of special interest in these communities, of course, are the buildings connected with mining: the gold refinery in Mariana, built between 1730 and 1760; Ouro Prêto's Casa dos Contos, one of the loveliest public buildings in the city; and in Sabará, the gold refinery (Casa de Fundicão e Intendência de Ouro) dating from 1750, which is today the Gold Museum.

Along with architecture, sculpture, and painting, music also flourished in Minas Gerais during the second half of the eighteenth century. Among the many singers, players, and composers were a number of mulattoes whose work was as thoroughly cosmopolitan as that of their "European" colleagues. The music is distinguished by remarkable technical skill. Most of the compositions were for voices, strings, wind instruments, and organ, and recordings are available of masses, hymns, and novenas by the eighteenth-century Mineiro composers José Joaquim Emerico Lobo de Mesquita, Marcos Coelho Neto, Francisco da Rocha, and Ignacio Parreira Neves.[25]

The attention paid to music by the well-to-do miners and religious brotherhoods which had money to sponsor composers and performers is of unusual historical interest. It is, in short, no accident that Brazil's great baroque outpouring in architecture, sculpture, painting, and music should have taken place in the mining province.

Turning from music and art to agriculture, the imaginative spirit of the Iberians is illustrated by the zeal with which they introduced useful plants into the New World, beginning with Columbus's second voyage. After a vast number of attempts and failures, almost all the species of cultivated plants known in Europe were acclimated and adapted to the quality and location of the soil in which they were planted.[26] Similarly, Spain and Portugal introduced American plants into Europe, Africa, and Asia, and the agricultural enrichment was equally revolutionary. Maize, manioc, white potatoes and yams, tomatoes, peanuts, cacao, pineapples, cayenne pepper, allspice, piñon nuts, cashew nuts, vanilla, and tobacco were all native to the Americas and had been unknown in Europe before the sixteenth century.

One of the most striking contributions to the new settlements was the attention paid to education, particularly by Spain. From the earliest period of colonization, members of monastic orders took charge of the evangelization and teaching of the Indians. They also realized that a knowledge of Indian languages was indispensable and set about the task of learning them. The Franciscans worked in Mexico City and the surrounding area, which constituted the central core of the country. The Augustinians labored in the province composed of the present-day states of Hidalgo and Michoacán; the Dominicans were assigned to Oaxaca. Four years after the occupation of Tenochtitlán–Mexico City, the padres were teaching the Indians the Gospel in their own languages and teaching them how to read and write in Spanish. Padre Sahagún and his companions even founded the Colegio of Tlatelolco in the capital, where Indians were prepared for the Catholic priesthood.[27] In Brazil the Jesuits labored for many years among the Tupí-Guaraní tribes, and among them the dedicated Joseph de Anchieta was remarkable for his plays and poems in the native idiom.[28] His grammar and vocabulary became the textbook in the Jesuit colleges of Brazil and was finally published at Coimbra in 1595. In the next century, Padre Vieira was active among the Indians of the north for many years, learning seven different languages, and as if that were not enough, he mastered Kinimbu in order to preach to the slaves arriving from Angola.[29]

The need for disseminating learning materials as efficiently as possible moved the first Archbishop of Mexico, Fray Juan de Zumárraga, to request the establishment of a printing press in Mexico. This press was installed in 1535, a mere fourteen years after the capture of the Aztec capital and a full hundred years before the first English-language printing press was set up in the Massachusetts Bay Colony. Some of the earliest books were printed in Mexican languages as well as in Spanish and dealt with Christian doctrine.[30] Madariaga notes:

> In the last quarter of the seventeenth century there were four printing presses in New Spain. Printing presses were already at work in Lima (1584) and Guatemala (1667). In 1761 there were six printing presses in Mexico City, one of which could print Greek and Hebrew. The press was introduced in Paraguay in 1705, in Santa Fé de Bogotá in 1739, in Quito in 1755, in Habana in 1765, in Buenos Aires in 1766, in Caracas in 1806, in Chile in 1812. As a term of comparison, Boston [Cambridge] had its first press in 1638 [actually 1636], Pennsylvania [Philadelphia] in 1686; New York in 1693; Virginia in 1729; Jamaica in 1756. As for England herself, Macaulay records that in 1685 'except in the capital and at the two universities, there was scarcely a printer in the Kingdom. The only press in England north of Trent appears to have been at York.' And he adds: 'There had been a great increase within a few years in the number of presses; and yet there were thirty-four counties in 1724 in which there was no printer, one of those counties being Lancashire.'[31]

In 1551 the first two universities in the New World were founded—San Marcos in Lima, the capital of the Viceroyalty of Peru, and that of Mexico in the capital of the Viceroyalty of New Spain. The former institution was established by a royal cedula signed by Queen Joanna on May 12, 1551, and the latter by a decree of Prince Philip on September 21. Although there is some uncertainty about which of these universities first opened its doors, at least we can establish by the year of their founding the importance that Spain attached to higher education in its newly organized territories.

The University of San Marcos at Lima came into being a bare sixteen years after the city itself; but the Dominicans had already been teachng subject matter suitable for general studies for three years in the location where the university was established. The University of Mexico was requested jointly by the Bishop, Fray Juan de Zumárraga, the city council, the religious orders, and the Viceroy Antonio de Mendoza.[32]

In the city of Chuquisaca (later called La Plata, Charcas, and finally Sucre, its present name), within the jurisdiction of Potosí, a university was founded in 1624 and named for St. Francis Xavier in accordance with instructions issued by King Philip III. As its fame grew, it attracted students from the entire region and from the provinces of Río de la Plata. The King later conferred higher status upon the university by granting it the same rights and privileges as those held by the University of Salamanca in Spain.

In the Viceroyalty of Río de la Plata, the University of Córdoba, the fourth in America, was established to cover the southern part of the vast territory controlled by Spain. Córdoba opened its doors in 1614, although the Papal Bull authorizing it did not arrive until 1621 and the royal cedula, granted by Philip III, until 1622.

In the eighteenth century, that century of enlightenment and scientific advancement, the progressive government of Charles III decided that for better exploitation of the mines of the Indies, it would be advisable to send German technicians to the King's overseas domain. These specialists were to observe and comment on the systems employed and to establish colleges for the preparation of mining technicians, or as we would say today, engineers. When these men graduated, they were to possess all the knowledge necessary to take complete charge of mining operations in the most practical and efficient way possible.

The task of inaugurating the Theoretical and Practical Academy and School of Metallurgy in 1779 fell to the very intelligent civil servant Don Jorge Escobedo, Governor of Potosí. At the official opening ceremony on February 3, he delivered a lengthy speech and read the academy's governing ordinances; these regulations were approved by the Crown on January 14, 1780. A short time later the Spanish government sent to Potosí a commission of German scientists headed by Baron Thaddeus von Nordenflicht, which carried out extensive mining and mineralogical studies.[33]

In New Spain there were several delays in opening the college provided for by the Mining Ordinances of 1783 for the technical preparation of men to direct mining operations.[34] As early as July 1786, Charles III had appointed a Director of the Royal Seminary to enable him to make the necessary preparations beforehand. The honor of the appointment went to Don Fausto de Elhuyar y Zubice, one of the most eminent scientists in the principal European centers of study. Humboldt said of him that he was "a famous savant in the annals of the chemical sciences." Elhuyar was born in Logroño, Spain, on October 11, 1755, and was educated in Paris by outstanding teachers. In 1777 he was a student at the Basque Seminary of Vergara and at the age of twenty-three received a scholarship to visit various European capitals. He studied with Abraham Gottlob Werner at the Academy of Chemnitz in Freiberg and then returned to Vergara in 1782. His chief fame arose from the fact that, in 1783, at the age of twenty-eight, while occupying the Chair of Chemistry and Mineralogy, he discovered and isolated the element tungsten in collaboration with his brother Juan José. After this, he again visited Chemnitz in Saxony to study the amalgamation technique of Baron von Born. Soon after his return to Spain, Elhuyar set out for Mexico in August 1788, accompanied by eleven German mining engineers.[35]

Among the professors at the Seminary of Mining, one superb teacher, scholar, and writer was preeminent: Don Andrés Manuel del Río. He arrived in Mexico in 1794 and, until his death in 1849, dedicated his long and intensive scientific life to his adopted country. Del Río was born in Madrid on November 10, 1764, and obtained a bachelor's degree from Alcalá de Henares in 1782. After teaching for a year at the Royal Academy of Mines at Almadén, he received a scholarship to study in Paris. He next went to Freiberg where he worked with Werner, and subsequently with Ruprecht, at the Academy of Chemnitz in the field of chemistry; Humboldt and von Buch were his fellow students. Back in Paris, he attended classes at the Laboratoire de L'Arsenal, where Lavoisier was director. The events of 1794 which led that great French chemist to the guillotine obliged del Río to leave for England where he studied metallurgy. Summoned to Mexico by Fausto de Elhuyar, he arrived there in December 1794, thirty years old and magnificently prepared to teach in the Royal Seminary. His labors in the classroom, in laboratories, and in the investigation of Mexican ores were extraordinary. He wrote a large number of books and discovered many mineral elements. Guyton de Merveau has described del Río's *Tratado de orictognosia (Treatise on Oryktognosie [Practical Mineralogy])* as the most notable work on mineralogy of its time in any language. The achievement which brought him the most fame was the discovery of vanadium, which he named erythronium. The noted Mexican historian Dr. Arturo Arnaiz y Freg, writing of del Río's life, stated:

> When at the end of 1827 Mexico decreed the expulsion of Spaniards, Andrés M. del Río headed the list of those excepted from this rule. However, the famous mineralogist wished to suffer the same fate as his countrymen and voluntarily went into exile. Hence he arrived in the United States, where for six

> years he received great honors in Washington, Philadelphia, and Boston. After his return to Mexico in 1835 he made every effort to forget the unjust expulsion of the Spaniards, and lost no time in feeling just as much a Mexican as he had before.[36]

Portugal as well as Spain felt that its mines in the New World required European-trained experts, but Lisbon differed in one respect from Madrid; its principal authorities were Brazilian-born specialists who, after completing courses at Coimbra, continued their studies in France, Germany, and other countries. The outstanding mineralogists were Manoel Ferreira da Câmara Bethencourt e Sá (1765–1835), José Bonifacio de Andrada e Silva (1765–1838), and, to a lesser extent, the latter's brother Martim Francisco Ribeiro de Andrada e Silva (1775–1844).

Câmara was born in the Captaincy of Minas Gerais and obtained his degree in Portugal with distinction in 1787. Two years later he won a prize from the Royal Academy of Sciences of Lisbon, and his first formal address before that body was entitled *Physical and Economic Observations on the Extraction of Gold in Brazil.* He pointed out that mining for gold in Minas Gerais was difficult because of the chemical content of the gritstone (hematite, magnetite, or ileminite) in the riverbed gravel. On the other hand, if one dug into the hillsides, the soil tended to be either extremely friable or very hard to tunnel.[37]

José Bonifacio was born in the seaport town of Santos and early showed remarkable gifts. He mastered half a dozen languages and ultimately was celebrated not only as a scientist but also as a jurist, a poet, and a statesman. Among his writings are treatises and speeches on geology, agronomy, government, abolition, and Indian acculturation, as well as a volume of neoclassic verse. After studying in Santos and São Paulo, he left at the age of twenty for Coimbra and in 1787 received degrees in natural philosophy (science) and law.[38]

One of the outstanding Portuguese ambassadors and statesmen at the time was D. Rodrigo de Sousa Coutinho, who in 1790 drew up a paper *On the True Influence of Mines of Precious Metals in the Industry of Nations That Possess Them and Especially Those of Portugal.*[39] It was he who chose the two budding scientists from overseas, together with a young Portuguese, to carry out advanced research in Europe for nearly a decade and appointed Câmara, who possessed administrative as well as scientific abilities, leader of the group. Letters of introduction were prepared for the Portuguese diplomatic representatives in the countries to be visited, and since the Foreign Minister, Luis Pinto de Sousa, Count of Balsemão, had been Captain-General of Minas Gerais from 1767 to 1772, he well understood the goals of the scholars. The first stop was Paris, where the young men, working in the school in which Lavoisier was director, followed courses in chemistry with Jean Baptiste Chaptal and Antoine François Fourcroy and with the Abbé Haüy in mineralogy. It was soon evident that the Brazilians were not novices, and José Bonifacio presented a *Mémoire sur les diamants du Brésil* before the Société d'Histoire Naturelle which appeared in the *Annales de chimie* in 1792.[40]

Like their Spanish counterparts, the youthful scientists also attended the mining academy at Freiberg in Saxony where they studied under the great geologist Abraham Gottlob Werner, the first modern scholar to classify minerals systematically. One of their fellow students was Alexander von Humboldt, and there they also met Baron Wilhelm von Eschwege, who was later to spend years in Brazil as a metallurgical adviser. The mineral deposits of the Tyrol, Steiermark, and Carinthia were next investigated, and a sojourn was made at the University of Padua where the group attended lectures by the distinguished physicist Alessandro Volta (after whom the *volt* is named). During these travels, Câmara was particularly interested in the practical side of mineralogy, and he worked variously as a draftsman, carpenter, brickmaker, mason, and smelter. Indeed he spent so much time beside hot furnaces and forges that he later said this may have damaged his health. José Bonifacio was more theoretically minded than his countryman. While studying with Volta, for instance, he wrote a detailed geological analysis of the Euganian hills between Padua and Este.[41] A trip to England included talks with the English scientist and theologian Joseph Priestley. One of the most interesting visits was to Scandinavia in 1797, where José Bonifacio again distinguished himself by discovering four new mineral species, criolite, petalite, scapolite, and spodumene, and also added to existing information on eight other varieties of these species. His findings were soon published at Paris in 1800 as *Exposé succinct des caractères et des propriétés de plusieurs minéraux de Suède et de la Norvège, avec quelques observations chimiques faites sur ces substances,* and in 1803 there followed *Notice sur la structure minéralogique de la contrée de Sala en Suède.*[42] Since he was obviously a superior geologist, José Bonifacio was offered the position of General Intendant of Mines in Norway by the Prince Royal of Denmark. He felt a responsibility to the Portuguese-speaking world, however, and returned to Coimbra to accept a professorship of metallurgy and geology (Geognosy) in 1800. Manoel Ferreira da Câmara had returned in 1798 and was immediately asked to advise the Crown on mining matters. He also made two important studies on currency and coining: *Memoir on the Acquisition and Coining of Copper* and *Memoir on the Changing of All Gold Dust into Currency.*[43] Another task, undertaken at the suggestion of D. Rodrigo, was to draw up new mining laws for Brazil to supersede those established in 1702, while taking into consideration the code which Pombal had devised in 1771. At about this time, the Portuguese government realized that its experts were ready to implement their findings and concluded that Câmara should be the new Intendant of Mines in Brazil while José Bonifacio would assume the same post in Portugal. Câmara's appointment was held up when D. Rodrigo resigned as Minister, but he sailed nonetheless for Brazil to look after his family affairs and to be the metallurgist of an undertaking in the field of iron production.

Câmara was the first in Brazil to make pig iron, at the Real Fábrica de Ferro do Morro do Pilar in Minas Gerais, and the development of the country's iron industry dates from his time. In 1807 he became Intendant of Mines with headquarters in Minas Gerais, a position he held with distinction until 1823. Another development he fostered was the increased production of saltpeter, which induced him to set up a gunpowder factory.

The younger brother of José Bonifacio, Martim Francisco, also devoted his energies to mineralogy, and in 1799, after graduating from Coimbra, he translated and published Farber's two-volume *Manual of Mineralogy.*[44] He was named Inspector of Mines in the Captaincy of São Paulo in 1802 and once back in Brazil traveled extensively throughout the interior. The results of his investigations are found in *Jornais das viagens de 1803 e 1804 (Journals of Trips in 1803 and 1804)* and *Diário de uma viagem mineralógica pela província de São Paulo (Diary of a Mineralogical Trip through the Province of São Paulo).*[45] José Bonifacio meanwhile continued to teach at Coimbra, act as Intendant of Mines in the Peninsula, and publish studies such as *Mémoria sôbre as minas de Portugal (Memoir on the Mines of Portugal)* and *Mémoria sôbre a nova mina de ouro da outra banda do Tejo, chamada Principe*

Regente (Memoir on the new gold mine on the other side of the Tagus, called the Prince Regent).[46] Meanwhile the Abbé Haüy in Paris, in recognition of his former pupil and colleague, gave the name *Andradita* to a black garnet, one of the eight species which José Bonifacio had described in Sweden. Both Câmara and José Bonifacio were members of the Royal Academies of Science in Lisbon, Stockholm, and Edinburgh, as well as other learned bodies, and José Bonifacio also belonged to the American Philosophical Society.

With the arrival of Dom João in 1808, both Martim Francisco as Intendant of São Paulo and Câmara as Intendant in Minas Gerais were able to work for improved mineral production, particularly as Dom Rodrigo de Sousa was now Chief Minister of the Crown and residing in Rio de Janeiro. The latter's death in 1812 was a blow to scientific progress in Portuguese America. Perhaps Câmara's most foresighted analysis as a mineralogist was his prediction that there was a much greater future for iron than for gold in Brazil. In 1823 the veteran Intendant drew up a memorandum on the need for founding an *academia montanistica,* but it was years before an escola de minas was finally established.[47]

Notes

1. The term *encomienda* referred originally to the allotment of Indian workers by the conquistadors to the colonists. Later the word *repartimiento* signified temporary allotment of Indians for a given task, while an encomienda eventually became a hereditary land grant [Bailey Diffie, *Latin American Civilization: Colonial Period* (Harrisburg, Pa.: Stackpole Sons, 1945), p. 61]. The conditions at first were that the Indians' subsistence and religious indoctrination were assured and their juridical freedom and the supreme sovereignty of the Crown were safeguarded. The *mita* was obligatory labor for a specified period of time on public works considered indispensable or urgent to the Crown or the community. This work might also include herding and mining. Both systems were abused in many ways, and elaborate means for evasion were contrived to offset the elaborate legal procedure set up to protect the Indian.

2. The distinguished Mexican engineer Don Luis A. Bracamontes, in a lecture delivered in 1968 as part of the commemoration of the one hundred and seventy-fifth anniversary of the founding of the College of Mining in Mexico City, made the following statement: "The Laws of the Indies contained a chapter dedicated to the location and design of towns, taking into account what today we would call socio-economic factors, and, in consequence, regulating the placement and dimensions of the town square, the church and municipal buildings, and the other construction

necessary for establishing a town or city, as the case might be. It may be said that this was the origin of urban engineering, and that, in fact, what is called city planning today, with its group of ordinances, was defined at that period [the sixteenth century], and that these ordinances constituted the first planning law in the world."

3. A great deal has been written on the fine arts of Latin America, and the bibliography of Robert C. Smith and Elizabeth Wilder, *A Guide to Latin American Art* (Hispanic Foundation, The Library of Congress, Washington, D.C., 1948), is indispensable for the earlier material. Among general histories, attention should be called to George Kubler and Martin Soria, *Art and Architecture in Spain and Portugal and in their American Dominions, 1500–1800* (London: Penguin Books, 1959); Pal Kelemen, *Baroque and Rococo in Latin America* (New York, 1951); Miguel Solá, *Historia del arte hispano-americano* (Barcelona, 1935); Angel Guido, *Descubrimiento de America en el arte* (Buenos Aires, 1944); Diego Angulo Iñiguez, *Historia del arte hispano-americano,* 3 vols. (Barcelona-Madrid, 1945–1956); Manuel Toussaint, *Arte mudéjar en América* (Mexico, 1945); and Pal Kelemen, *Art of the Americas, Ancient and Hispanic* (New York, 1969).

4. Sylvester Baxter, *Spanish-Colonial Architecture in Mexico,* 12 vols. (Boston, 1901), vol. 1, p. 196. A Spanish translation of the text with less elaborate illustrations but with excellent notes by Manuel Toussaint was published as *La arquitectura hispano-colonial en México* (Mexico, 1934).

5. For a survey of Mexican art see Manuel Toussaint, *Colonial Art in Mexico,* Elizabeth Wilder Weismann trans. and ed. (Austin and London, 1967), and Pedro Rojas, *Historia general del arte mexicano: Era colonial* (Mexico, 1963).

6. See Antonio Cortés, *Valenciana: Guanajuato, Mexico* (Mexico Secretaría de Educación Pública, 1933); Nicolau Armando, *Valenciana* (Mexico: Dirección de Monumentos Coloniales, Instituto Nacional de Antropología e Historia, 1961); and Baxter, op. cit.

7. Manuel Toussaint, *Tasco: Su historia, sus monumentos* (Mexico City, 1931); Manuel Toussaint, *Guia illustrada de Tasco . . . con traducción inglesa [with English translation]* (Mexico City, 1935); and Baxter, op. cit., pp. 193–195.

8. Francisco de la Maza, "El arte en la ciudad de Nuestra Señora de los Zacatecas," *Mexico en el arte,* vol. 7 (1949), pp. 5–16, and José R. Benitez, *Las catedrales de Oaxaca, Morelia y Zacatecas* (Mexico City: Talleres Gráficos de la Nación, 1934).

9. Manuel Muro, *Historia de San Luis Potosí* (San Luis Potosí, 1910).

10. See Benitez, op. cit.; Antonio Vázquez de Espinosa (d. 1630), *Descripciónes de la Nueva España en el siglo XVII* (Mexico City, 1944),

p. 147, and Alfonso Francisco Ramirez, *Hombres notables y monumentos coloniales de Oaxaca* (Mexico City, 1948), p. 168 ff.

11. Justino Fernández, *El Palacio de Minería* (Mexico City: Instituto de Investigaciones Estéticas, Universidad Nacional Autónoma de México, 1951), and Kubler, op. cit. p. 82.

12. Pal Kelemen, *Latin American Architecture,* p. 210. Kubler says that this altar is the work of Vicente Javier who was active between 1748–1780, op. cit. p. 171.

13. Kubler, op. cit., p. 89.

14. Santiago Sebastián, *Arquitectura colonial en Popayán* . . . (Cali, 1965), p. 16.

15. Kubler, op. cit., p. 97.

16. See Martin Noel, *Las iglesias de Potosí: Documentos de arte colonial sudamericano, cuaderno 3* (Buenos Aires, 1945); H. E. Wethey, "La última fase de la arquitectura colonial en Cochabamba, Sucre y Potosí," *Arte de America y Filipinas,* (Sẽville, 1952), vol. II, p. 4; idem, "Hispanic Colonial Architecture in Bolivia," *Gazette des Beaux Arts* vols. 39–40 (New York, 1952), pp. 47–60 and pp. 193–208. The architectural section in Angulo Iñiguez,op. cit., was written by Enrique Marco Dorta. For San Lorenzo, consult vol. III, pp. 516–519.

17. Angulo Iñiguez, op. cit., vol. III, pp. 532–534. See also Manuel Toussaint, "La Casa de Moneda de Potosí," *Boletín de la Sociedad Geográfica* (Potosí, 1940).

18. Angulo Iñiguez, op. cit., vol. III, pp. 493–497.

19. Martin Soria in George Kubler, op. cit., pp. 325–326. See also Jose de Mesa and Teresa Gisbert de Mesa, "Un pintor colonial boliviano: Melchor Pérez Holguín," *Arte de America y Filipinas,* vol. II, no. 4 (Seville, 1952), pp. 149–216.

20. Robert Stevenson, *Music in Mexico: A Historical Survey* (New York, 1952), pp. 135–136.

21. Stevenson, op. cit., pp. 149–153.

22. The record *Salve Regina, Choral Music of the Spanish New World 1550–1750* was made by the Roger Wagner Chorale for Angel Records. It is no. 36008.

23. Robert Stevenson, *The Music of Peru: Aboriginal and Viceroyal* (Washington, 1960). For Gutiere Fernandez Hidalgo, see pp. 182–184; for Juan de Araujo, see pp. 187–190.

24. There is an extensive literature on Brazilian eighteenth-century art, and Minas Gerais was of course a creation of that period. Thanks to the Patrimônio Histórico e Artístico Nacional, much of it has been preserved. The following paragraphs are based on Edgard de Cerqueira Falção, *Relíquias da terra de ouro* (São Paulo, 1946); and Germain Bazin, *L'architecture religieuse baroque au Brésil* (São Paulo and Paris, 1957); Germain Bazin, *Aleijadinho et la sculpture au Brésil* (Paris, 1963); and Dom Clemente Maria da Silva-Nigra, "Artistas coloniais Mineiros," *Revista de História,* vol. 2, no. 6 (April–June 1951).

25. The leading authority on early Brazilian music is the Uruguayan scholar Francisco Curt Lange who has written extensively on the subject. His chapter, "A Música Barroca," in Sergio Buarque de Holanda's *História geral da civilização Brasileira: A época colonial,* 2d ed. (São Paulo, 1963), part I, vol. 2, pp. 121–144, gives an excellent summary of music in the mining district. A statement from a slightly irritated judge of the Court of Appeals in 1780 to the Crown is revealing: "There are so many musicians in the Captaincy of Minas that they certainly exceed the number of those in all Portugal." The recordings *Mestres do Barrocco Mineiro, século XVIII,* vols. I and II, were made by the Associação de Canto Coral de Rio de Janeiro and the Orquestra Sinfônica Brasileira under the direction of Edoardo de Guarnieri, nos. 5005 and 5006.

26. The great Americanist Don José Tudela lists the following plants as introduced into the New World in his study *El legado de Espana a América,* 2 vols. (Madrid: Ediciones Pegaso, 1954), vol. II:

    Seeds: 1. Cereals: wheat, barley, rye, oats, rice
    2. Legumes: lentils, peas, chickpeas

    Green vegetables: Lettuce, escarole, chard, saltwort, cabbage, cauliflower, kale, onions, scallions, leeks, asparagus, artichokes, celery, sesame, borage, spinach

    Root vegetables: Radishes, turnips, beets, carrots

    Fruits: Squash, cucumbers, citron, watermelons, melons, eggplant, oranges, lemons, grapefruit, limes, apples, pears, quinces, peaches, apricots, "Paraguay peaches" (despite their name), cherries both tart and sweet, pomegranates, figs, large and small strawberries, raspberries, bananas (a few varieties were native, but others came from the Canary Islands and Africa), and the Philippine mango.

    Nuts: Walnuts, almonds, and hazelnuts

    Spices and seasonings: Saffron, anise, garlic, parsley, cummin, laurel, ginger (brought not from Asia but from Spain).

    Other plants, fruits, and their products: Grapes and wine, olives and olive oil, sugar cane, and coffee whose introduction into America is disputed by the Dutch and French but whose use was spread by the Spaniards.

Textile plants and fodders: Flax, hemp, willow, alfalfa, clover

It is touching to read in Bernal Díaz del Castillo's account of the Conquest of Mexico how he carried in his pack some orange seeds which he says were planted in New Spain for the first time when he accompanied Grijalva's expedition. See also James Robertson, "Some Notes on the Transfer by Spain of Plants and Animals to its Colonies Overseas," *James Sprunt Historical Studies,* vol. 19, no. 2, (Chapel Hill: University of North Carolina Press, 1927), pp. 7–21.

27. It is to Fray Bernardo de Sahagún that we are most indebted for information about the indigenous cultures. In his great work *Historia general de las cosas de la Nueva España (General History of the Things of New Spain,* 1560) on which he spent forty years, Sahagún wrote that expert indigenous students were his collaborators and that they knew Nahuatl, Spanish, and Latin. In Robert Ricard, *The Spiritual Conquest of Mexico,* Lesley Byrd Simpson, trans. (Berkeley: University of California Press, 1966), there is mention of a considerable number of missionary priests who were skillful linguists and who, from the earliest years of the occupation of Mexico, knew, spoke, and wrote a number of indigenous languages as an indispensable aid to evangelization and teaching.

28. See *Cartas, informações, fragmentos históricos e sermões de Padre Joseph de Anchieta, S. J.* (1554–1594) (Rio de Janeiro: Academia Brasileira de Letras, 1933) and José de Anchieta, *Arte da grammática da lingoa mais usada no costa do Brasil* (Coimbra, 1595). A facsimile edition was published by the Imprensa Nacional (Rio de Janeiro, 1933).

29. Vieira was noted for his defense of oppressed peoples. His famous sermon delivered on the first Sunday of Lent, 1653, violently condemned Indian slavery and the inhabitants of São Luis de Maranhão who sent out expeditions to capture natives to work their plantations and recalls the vehemence of Padre Montesinos and Bartolomé de las Casas. See Bradford Burns, *A Documentary History of Brazil* (New York, 1966), pp. 82–89, for a translation of parts of the sermon. The complete text is found in Hernâni Cidade, *Padre Antônio Vieira* (Lisbon, 1940), vol. III, pp. 157–186.

30. As early as 1533, Fray Juan de Zumárraga was making arrangements with the Council of the Indies for a printing press in Mexico. The first book printed in Mexico is thought to be the *Doctrina Cristiana en lengua mexicana y castellana (Christian Doctrine in the Mexican and Castilian languages)* printed by Juan Pablos in 1539, but there are many indications that by 1535 a printer named Esteban Martín, who certainly lived in Mexico at about that date, had already edited the *Escala espiritual de San Juan Clímaco.* The classic study of this subject is José Toribio Medina, *La imprenta en México,*

1539–1821, 8 vols. (Santiago de Chile: Impreso en Casa del Autor, 1907–1912).

31. Salvador de Madariaga, *Rise of the Spanish American Empire* (New York, 1947), pp. 39–40.

32. Pope Pius V dispatched the Bull of July 1571, confirming the institution's existence and granting it the distinctions and privileges requested by Charles V. The first Rector of the University of San Marcos was the Dominican friar Juan Bautista de la Boca. This initial stage of church control was replaced on May 11, 1571, by a new order of things, for the Audiencia authorized the professors to choose a lay rector freely. They elected Dr. Pedro Fernández Valenzuela, who instituted the secular period of the university, warmly supported by the Viceroy, Don Francisco de Toledo.

    The professors who founded the University of Mexico were famous scholars including Fray Alonso de la Veracruz and Juan Negrete, Master of Arts of the University of Paris. In 1595, Philip II obtained from Pope Clement VIII the pontifical sanction which conferred the protection of the university code, universal recognition of its course of study, and, upon its graduates, the *jus ubique docendi,* or right to teach everywhere. For three centuries the Illustrious, Royal, and Pontifical University of Mexico was, according to Dr. John Tate Lanning, the most advanced cultural center in the New World, as well as the pride of the citizens of New Spain.

33. Marius André, "Le Baron de Nordenflicht . . . et les mineurs allemands au Pérou," *Revue de l'Amérique latine,* vol. VIII (1924), pp. 289–306; Carlos Denusta Pimentel, "La expedición mineralogista del Baron Nordenflicht al Perú," *Mercurio Peruano* (Lima), vol. XXXVIII (October–November 1957), pp. 510–519.

34. While study plans were being prepared, the Royal Tribunal of Mining was arranging for the necessary funds, in accordance with the provisions that had been made. The college was finally inaugurated on January 1, 1792, in the Viceroyship of the second Count of Revilla Gigedo. The subject matter for the four-year course of study was the following: in the first year, arithmetic, algebra, elementary geometry, plane trigonometry, and conic sections; in the second year, practical geometry applied to typical mining operations (including subterranean geometry), and in addition dynamics and hydrodynamics; in the third year, chemistry, restricted to the mineral kingdom, including recognition of minerals, their constituent principles, and the methods employed to analyze them, as well as metallurgy, or treatment of the various methods and operations generally used to extract all subterranean products; in the fourth year, subterranean physics or theory of mountains, intended to serve as an introduction to the subsequent study of mining work and the tasks required for underground excavations "from the first inspection of the terrain to the extraction of its products and other materials." There were

also to be classes in drafting and French. After finishing the third and fourth years, the students were expected to spend two or three months in the mines nearest the capital working under the supervision of their teachers in order to gain practical experience in what they had been learning in theory. In addition to the necessary classrooms, the seminary was to have two display rooms: one for models, machines, kilns, and various implements; the other for minerals and the products of their extraction. There was also to be a laboratory of chemistry. [Dr. José Joaquín Izquierdo, *La primera casa de las ciencias en México: El Real Seminario de Minería, 1792–1811* (Mexico: Ediciones Ciencia, 1958).]

35. When Elhuyar returned to Spain thirty-three years later (1821), he was appointed Director General of Mines. He died on January 6, 1833, in Madrid, as the result of an accident.

36. See the biographies of Don Andrés M. del Río by Santiago Ramírez and by Arturo Arnaiz y Freg; also the observations in Modesto Bargalló, *La mineria y la metalurgia en la América Española durante la epóca colonial* (Mexico, 1955), and Izquierdo, *La primera casa de las ciencias en México.* The periodical *Ciencia,* of Mexico City dedicated vol. XXII, no. 5 (1964) to the commemoration of the second centenary of Andrés Manuel del Río's birth.

37. The classic study on Manoel Ferreira da Câmara Bethencourt e Sá is Marcos Carneiro de Mendonça *O Intendente Câmara,* 2d ed. (São Paulo, 1958). A prize was given for his *Ensaio de descrição física económica da comarca dos ilheus na América* which appeared in *Mémórias económicas da Academia Real das Ciências de Lisboa* (Lisbon, 1789), tomo I, pp. 304–350. He also prepared at this time *Observações acerca do carvão de pedra da Freguesia da Carvoeira.* Câmara was a practical scientist who understood the relationship of mineralogy to industry and agriculture. For his comprehension of the importance of cacao see Eusinio Lavigne, *Cultura e regionalismo cacaueiro: A personalidade de Manoel Ferreira da Câmara Bethencourt e Sá* (Rio de Janeiro, 1967).

38. The best biography of the great scientist and patriot is Otávio Tarquinio de Sousa, *José Bonifacio: 1763–1838* (Rio de Janeiro, 1945). In 1963 the Instituto histórico e geográfico brasileiro devoted the July–September number of its *Revista* to the father of Brazilian Independence, vol. 260 (Julho–Set.), pp. 153–335. It contains an article by Mario Barata, "Viagens de estudos científicos de José Bonifacio e atividades na intendência das Minas de Portugal," pp. 238–257. Pertinent information is also found in Elysario Távora, *José Bonifacio cientista, professor e técnico* (Rio de Janeiro, 1944), and Moses Bensabat Amzalak, *José Bonifacio de Andrada e Silva, economista* (Lisbon, 1941). This includes his "Memória sobre as minas em Portugal," pp. 19–41.

39. Dom Rodrigo de Sousa Coutinho's paper was entitled *Discurso sôbre a verdadeira influência das minas dos metaes preciosos na indústria das nações que as possúem e especialmente da Portuguesa.* For a general account

of his life see the Marquês do Funchal, *O Conde de Linhares* (Lisbon, 1908).

40. The French original and an English translation of Bonifacio's *Mémoire* are now available in Edgard de Cerqueira Falcão, ed., *Obras cientificas, politicas e sociais de José Bonifacio de Andrada e Silva* (São Paulo, 1965), pp. 57–60.

41. The study of the Euganian hills was originally made in 1794 but was only published as *Viagem geognostica aos montes Eugâneos, no território de Padua, na Italia* in the *Memórias da Academia Real das Sciencias de Lisboa* in 1812.

42. The contemporary German and English versions of the "Exposé" have been reprinted in Cerqueira Falcão, op. cit., pp. 61–73 and pp. 85–93. The *Notice sur la structure minéralogique* is found in *Journal des Mines,* vol. XV (Année XII, 1803–1804), pp. 249–259.

43. The *Memória sobre a acquisacão e cunhagem do cobre* written in 1799 is reprinted in Mendonça, op. cit., pp. 76–83, and the *Memória sobre a permuta de todo o ouro em pó por moeda corrente,* ibid., pp. 295–303.

44. Antônio Carlos's translation was entitled *Manual de mineralogia, ou esboço do reino mineral, disposto segundo a analyse chimica de Mr. Farber, . . .* 2 vols. with engravings (Lisbon, 1799). In 1800 he also made a study trip with his brother in Portuguese Extremadura and Beira, and a report of it was published soon after by José Bonifacio in the *Freiberg Journal of Mines.*

45. The journals of 1803 and 1804 are found in the *Revista do Instituto Brasileiro Histórico,* 1847, pp. 537 ff.

46. The *Memória* on the mines of Portugal appeared in *O investigador Portuguêz em Inglaterra* (London, 1814), vols. 10 and 11, nos. XL, XLI, and XLII. The *Memória sôbre a nova mina de ouro . . .* came out in *Histórias e Memórias da Academia Real das Sciencias de Lisboa* tome V, part I, pp. 140–152. 1817.

47. One of Câmara's ideas was to build a canal using the water of the Santo Antônio River to connect with the Doce River and thus send iron ore cheaply to the coast.

XI

The Beginning of a New Era

So far this study has reviewed the process of the formation and development of the American nations through mining and economic activity. But miners were destined to play yet another decisive and dramatic role in America's history. This was their part in the insurgent movements which culminated in the political emancipation of the American continent and the end of the era of the paramount importance of mining.

The first attempt to secure political independence in the New World took place in Portuguese America in Minas Gerais. Known as the *Inconfidência Mineira,* this revolutionary movement arose after a new tax was drafted in Lisbon in 1760. In the second half of the century, the province had been assessed 100 arrôbas of gold a year (13,300 pounds) instead of the customary Royal Fifth. However, the full sum had not generally been paid, and the Crown accused the Mineiros of fraud and contraband. When a special levy of 354 arrôbas of gold was proposed to cover the deficit for the period from 1774 to 1785, relations with the Crown became extemely tense. The indignation of the inhabitants was reflected in the thoughts of some of the community leaders and "enlightened" intellectuals who believed that the solution of the problem was to overthrow the government.

In 1786, José Joaquim da Maia, one of a group of young Brazilians studying at the University of Montpellier in France, wrote to Thomas Jefferson, then American Minister in France, mentioning a projected revolution and inquiring about the possibility of American aid.[1] Jefferson corresponded with Maia, and they met in the Roman amphitheatre of Nîmes in the spring of 1787. The North

American was diplomatically cautious and pointed out that he could not commit himself or his government, which had only recently been established. At the same time he expressed great sympathy for the concept of independence and implied that volunteers from the United States might come to the aid of Brazil.[2]

Another active member of this group was José Alvares Maciel, who had studied mineralogy in France and England and had brought back news of the Enlightenment. Maciel's brother-in-law was Commander of the provincial regiment of dragoons, which included a Lieutenant Joaquim José da Silva Xavier who avidly accepted the new ideas. Silva Xavier, an army engineer with an excellent knowledge of mineralogy, had also tried his hand at medicine and dentisty, which accounts for his nickname *Tiradentes* (Toothpuller). He attracted a group of Mineiros, many of them intellectuals with literary leanings (hence the phrase "Arcady and the Rights of Man"[3]), and in 1789 they discussed a plan to improve the economic situation of the mining district, adopt a rational tax system, encourage manufacturing, establish a university, modify the social order, emancipate the slaves, and set up a government independent from Lisbon. Among the leaders were Claudio Manuel da Costa, a lawyer, poet, and wealthy mine entrepreneur, and Tomás Antônio Gonzaga, a judge of the Court of Appeals whose lyric poetry is second only to Camões's in popularity.[4] Unfortunately, Tiradentes talked too much and the scheme was betrayed. The idealistic Lieutenant, learning that ten others had also been condemned to death, nobly went to the scaffold trying to assume sole blame for the projected uprising: "I am the cause of the death of these men. I wish I had ten more lives to give for all of them. If God will hear me, I alone shall die and not they." Actually, this is what happened, for the death sentences for the others were commuted to exile in Portugal's African and Asian colonies.[5] Although the Inconfidência Mineira failed, Brazilians today regard the outburst in the mining region as the first blow in their struggle for independence.

Brazilian independence was finally achieved through a movement led by José Bonifacio de Andrada e Silva who, as mentioned previously, had served as General Intendant of Mines in Portugal. Shortly after Napoleon's troops crossed the border from Spain in 1807, the Brazilian, who then held the Chair of Geology in Coimbra, was appointed Lieutenant Colonel of the Academic Battalion

Alexander von Humboldt, one of the foremost exponents of the importance of mining in the development of civilization, visited Spanish America from 1799 to 1804. His descriptions of mining activities in Mexico still provide the finest picture we have of Spanish American achievements in this field. The portrait, by Rafael Jimeno, was painted during Humboldt's stay in Mexico. (*Palacio de Minería de México.*)

of the University which fought with Wellington in the Peninsula campaign. When the war was over, Dom João, who had arrived in Rio de Janeiro with the Court early in 1808, raised Brazil to the status of a kingdom (December 1815), and as time went on, he seemed more and more inclined to remain in the overseas half of the dual monarchy. José Bonifacio therefore decided at long last to return to his native land. Several months after his arrival in 1819, and following conversations with Baron Eschwege, he and his brother Martim Francisco made a study of the minerals of the state of São Paulo, and their findings bore the title *Excursão montanistica da Província de São Paulo para determinar os seus terrenos metallíferos.*[6]

José Bonifacio was too interested in his country's future to restrict his interest to mineralogy, however, and with his two brothers, Martim Francisco and Antônio Carlos, began to take an active part in the Independence Movement. Similarly the Intendant Câmara found himself being drawn into the turn of events. Although lacking the political fervor of the Andradas, he had a strong sense of civic responsibility and became a member of the Provisional Government of Minas Gerais which was called to consider the Portuguese constitutional uprising of 1820. He was also one of the delegates who met in Rio de Janeiro in 1821 to debate Dom João's return to Lisbon and was pleased when Prince Dom Pedro stayed behind as Regent of the Kingdom of Brazil. The following year, after independence had been declared, he was a Deputy to the General Constituent Assembly and for a time its Vice President and then President. Finally, Câmara was a member of the committee chosen to draw up the Imperial Brazilian Constitution.[7]

Like Câmara, José Bonifacio was active in his province, São Paulo, and when the issue of constitutionalism arose in Portugal, he entered the Provincial Junta and by June 1821 was its Vice President and dominant figure, his brother Martim Francisco being Secretary. From that moment José Bonifacio was the guiding spirit of the Independence Movement. One of his first moves was to enlist the aid of Dona Leopoldina, the scientifically minded daughter of the Austrian Emperor Francis I, to encourage her husband Prince Pedro to throw in his lot with Brazil and not to return to Portugal. The majority of the country's Freemasons at the time were constitutional monarchists and, as Grand Master of the Brazilian Masonic Federation, José Bonifacio was able to work for a controlled monarchical government assisted by a disciplined organization. On January 9, 1822, Dom Pedro made the momentous decision, declaring: "Since it is for the good of all and for the general happiness of the Nation, I am ready . . . I shall remain." The Portuguese Cabinet of the Prince Regent thereupon resigned, and José Bonifacio became both Minister of the Realm and Foreign Minister. He remained at the head of the government during the next eighteen months, including the crucial days when independence was at last declared (September 7, 1822), the period of the coronation, and the organization of the Ministries. A falling-out with Dom Pedro over the drafting of the Constitution led to his return to Europe, but he was and is today rightfully regarded as the Father of His Country. On his tombstone in Santos is the inscription:

> With science he fulfilled the obligation which enlightened reason owed to nature. With action he paid the debt a citizen owed his country.[8]

Independence in Spanish America also was inevitable, a natural consequence of the growth and maturation of the peoples who formed the Spanish Empire. It was unfortunate, however, that events in the mother country, the occupation of Spain by French troops and the abdication of the King in favor of Napoleon, caused consternation, confusion, and contradictory feelings in America during those trying days, and that the pursuit of ill-fated, unwise policies by Ferdinand VII upon his restoration created the conditions that led to a frightful civil war. And this is precisely what the so-called War for Independence was. It was not a struggle between peoples differing in language, habits, and culture, but between men of comparable characteristics—"European" Spaniards and "American" Spaniards, mixed-bloods and mixed-bloods, Creoles and Creoles—divided on the one side by desire to preserve traditional systems and institutions and on the other to live in accordance with the new ideas of liberty and independence, or, at least, of political and administrative autonomy. These new desires had been consecrated by the Declaration of Independence of the United States in 1776, spread by the French Revolution, and registered by the Spanish Constitution of Cádiz in 1812. Deputies representing the American provinces or territories had participated in drawing up this Constitution, which was first sworn to and later repudiated by Ferdinand VII.[9] The battle of Ayacucho, fought in 1824 on a plateau in the silver-mining region of Southern Peru, gave the victory to the Independence Movement.

With this battle the war was over and most Americans and Iberians alike were united in the same feelings and aspirations. It is significant that once independence was achieved, only a few hundred people retired to the Peninsula, most of them functionaries and civil servants.

Some relevant and eloquent examples of the confusion of ideas and feelings aroused in the New World by events in the Peninsula support the thesis that the War of South American Independence was a real civil war. The officer corps of the diminished army that Spain maintained in its far-flung American territories was made up of Creoles, or "American" Spaniards, and the majority of the insurgent chiefs also were Creoles. Javier Mina the Younger, a Navarrese, provides an example. At the age of twenty, he was a brave guerrilla fighter harrying the French in Navarre, Aragón, and Catalonia, but after Spain was freed from French occupation, he went to Mexico to help the insurgents against the absolutist government of Ferdinand VII. On the South American continent, General José San Martín, born of Spanish parents in Las Misiones on the banks of the Uruguay River, lived in Spain from his seventh to his thirty-fourth year. He graduated from the Military Academy in Spain, and in 1812, after having acquitted himself with distinction in the Battle of Bailén where he was promoted to Lieutenant Colonel, he decided to return to Buenos Aires to help the Americans of his native country obtain their liberty. He accomplished his objective so successfully that he is known as the Liberator of Argentina, Chile, and Peru.

Agustín de Iturbide, author of the Plan of Iguala which proclaimed the Independence of Mexico in 1821, served in the royalist army up to that date. Viceroy O'Donojú, who accepted that plan in the name of Spain by the Treaty of Córdova, was deposed by the government of Ferdinand VII for having done so. The troops of the "terrible" Asturian, General Boves, the scourge of Bolívar in Venezuela, were composed of indigenous plainsmen from the Llanos of the Orinoco. And lastly, let us recall a poignant scene that took place on December 10, 1824, a few hours before the battle of Ayacucho, which ended Spanish control in America. An eyewitness account of the battle, quoted by Salvador de Madariaga in his

Bolívar, confirms again the fact that the American war of emancipation was a civil war.

> At eight o'clock General Monet, a strong, smart, bearded man, walked down to our lines, called Córdova and told him that, as there were several Spanish officers who had brothers, relatives and friends in the republican camp, he came to ask whether they could not meet before the battle. Córdova consulted Sucre, who gave his consent. For about half an hour, about fifty men on each side, leaving their swords at the line, conversed in a neutral space between the two camps. The Spanish Brigadier Tur, a young tall man of thirty-four, perhaps the one who had asked for the interview, ran to ask for his brother, a lieutenant colonel in the Peruvian army, about six years younger. "Ah, my brother, how sorry I am to see you covered with ignominy"—he cried out; and his brother, turning his back on him, retorted: "I did not come here to be insulted." But the Spaniard ran after him, and throwing his arms around his neck both wept for a long time.[10]

The ideological debate and confrontation between two famous South American jurists, the Paraguyan-born Don Pedro Vicente Cañete y Domínguez and the Aragonese-born Victoriano de Villalba, in regard to the events of 1810 also are illuminating in view of the quality and prominence of the opponents. Cañete was General Advisor of the Viceroyalty of Buenos Aires, of the Captaincy-General of Paraguay, and Magistrate of the government of Potosí. A man of vast culture, he was an upright administrator of public welfare, an indefatigable worker, and a diligent historian of the city and mines of Potosí. He was, also, a zealous defender of that much-debated institution—the mita—which established forced labor by Indians in the mines, a system he considered as much a bulwark of the Spanish Empire as military service for the survival of the Spanish monarchy in the Indies. Cañete remained loyal to the royalist side, risking loss of the fame and honor he had earned during years of faithful and efficient service, and even his health and his life.

His ideological opponent in the debate, Victoriano de Villalba, was professor in the University of Huesca, Prosecutor of the Audiencia of Charcas, Protector of Natives, and during his stay in Buenos Aires, Examining Judge of the administrative acts of the ex-Viceroy, the Marquis of Loreto. Two men, two subjects of the

King, two Jurists, two impassioned partisans, two irreproachable public servants—but they occupied two irreconcilable positions. The American was a staunch royalist, a partisan of systems of Indian labor "so that the economy of the State and of private persons will not be undermined." The Spaniard was "an immovable fighter for an ideal of justice, on the side of American beliefs."[11]

In view of their cultural conditioning, their technical knowledge, and their experience, it was natural that the engineers and others who had come out of the College of Mining should have embraced the new ideas and been active in the Emancipation Movement. It was they who handled the bronze and iron; who cast cannon; who manufactured gunpowder, munitions, and other arms; and who took over the minting of money. In every region of America, miners made valuable contributions. But in New Spain their participation assumed an extremely dramatic quality, as dramatic as that which characterized the actions of the very leaders of the insurrection. These men experienced neither glory nor triumph, for unlike their companions in South America, almost all the insurgent miners in New Spain died during the early days of the Independence Movement. This fact fully justifies the general respect and admiration in which they are held today.

In the case of Mexico, four of the most brilliant students of the Royal Seminary of Mining—all worthy pupils of Andrés M. del Río—committed themselves to the insurgent cause. All four were working in the mines of Guanajuato and immediately joined the workers of the Rayas and La Valenciana mines in support of Padre Hidalgo, the ex-Rector of the Seminary of Valladolid (modern Morelia), who had raised the cry of battle for independence in the nearby parish of Dolores on September 15, 1810. These young men were Casimiro Chovell, a miner's son from Tasco, with whom Humboldt had had some fruitful conversations and who was administrator of La Valenciana; José Mariano Jiménez, who after having worked in Sombrerete and Zacatecas was at the time employed in the mine of the Marquis de Rayas; Rafael Dávalos, a mathematics teacher in Guanajuato, who for several months in 1803 had helped Baron von Humboldt make geographical charts for which Humboldt had made public profession of gratitude and admiration; and Ramón Fabié, a native of Manila, who after having finished his studies at the Royal Seminary of Mining was gaining practical experience with Chovell in Guanajuato.

To this group we must add the name of another engineer, Isidor Vicente Valencia, a descendant of miners of Tlalpujahua (in the modern state of Michoacán). Valencia was also a student at the Royal Seminary and greatly admired by del Río. He had joined the insurgents in Zacatecas, where he was working when Hidalgo and his supporters passed through in defeat on their way North.

All these men lost their lives tragically, almost as soon as they joined the Movement. Chovell and Fabié were hanged on the gallows erected before the main door of La Alhóndiga de Granaditas (the Granary) in Guanajuato, the scene of terrible massacres of Spaniards two months previously. Dávalos was shot in the courtyard of the same building. Jiménez was captured and imprisoned with Father Hidalgo and Ignacio Allende and their companions in Acatita de Baján as they were fleeing toward the United States; he was shot in Chihuahua. And Valencia, though he had not participated in any act of war, also faced a firing squad.

Don Andrés M. del Río, who had so much admired these brilliant students and was so much grieved by their tragic deaths, always honored their memory. He even dedicated minerals he had discovered to both Chovell and Valencia. The name of *Chovelia* was given to a silicate of aluminum to honor, in del Río's words, "a hero of our nation and of mineralogy," and *Valencita,* an iodide of silver, honors "a distinguished student of mining whose deeds will perpetuate his memory."[12]

The political break with Spain and Portugal led to the establishment of twenty republics where today 211 million people speak Spanish and nearly half that number speak Portuguese. What had started out as risings to support Ferdinand VII when Spain was invaded by Napoleon turned into rebellion when Ferdinand's policies ignored the feelings of his New World subjects aware of the ideas of the Enlightenment and the principles of the North American and French Revolutions. Such was the wheel of fortune, even though many of the patriots were basically monarchists, including Miranda, Belgrano, San Martín, and for a time even Bolívar. The military and political role of Bolívar, the great Venezuelan soldier and statesman, is known to all. However, one detail regarding his faith in mining and his personal life should be recalled. When the worn-out liberator withdrew from the scene in 1830, intending to spend his final years in England, he arranged for the only property remaining to him, the Aroa mines in the province of Carabobo, to be sold to a group of Englishmen: "I have nothing else in the world on which to live or with which to pay my obligations."

In 1822 Portugal's Prince Regent, Dom Pedro, became the head of a constitutional monarchy which lasted until 1889, and it was José Bonifacio who had the vision and ability to guide the relatively peaceful separation from Portugal. Like Bolívar, the Brazilian statesman was not destined to end his days as the country's supreme civil authority or Minister of the realm. He was, however, entrusted for several years with the education of the youthful Dom Pedro II, who at an early age acquired a devotion to science unsurpassed by that of any other nineteenth-century Monarch.

The contributions of Spain and Portugal to the history of the Western Hemisphere are of epic proportions. The story of settlement and expansion has been recorded from the days of Columbus, and praise and blame have been heard on both sides of the Atlantic. In the final balance, however, the achievement is glorious. Neither mother country fully understood the problems of emancipation, but after the former possessions had blossomed into a number of youthful nations, the old ties took on new meaning and other dimensions. Spaniards and Portuguese continued to emigrate to the New World where they multiplied and prospered. The relationship of Latin America to the Iberian Peninsula persists in numerous ways, and even though the fact is often overlooked, the civilization which has evolved in the southern part of the hemisphere is an extension of the one which has existed over the centuries in Spain and Portugal.[13]

Notes

1. Jefferson's lengthy report of May 4, 1787, to the Secretary of State, John Jay, was more outgoing than his remarks at the interview, for as the Brazilian historian Francisco Adolfo de Varnhagen (son of the mineralogist) noted: "Fine diplomat that he was, [Jefferson] was cleverly concealing in the presence of this youth his own enthusiasm for such ideas." Pedro Calmon, 4th ed. *História geral do Brasil,* vol. IV, p. 399. See also Samuel Putnam, "Jefferson and the Young Brazilians in France," *Science and Society,* vol. X, no. 2, (Spring 1946), pp. 185–192.

2. Actually Captain David Jewett of New London, Connecticut, was second in command to Lord Cochrane when the few Portuguese troops and sympathizers were defeated by naval action in 1823. There were of course many more English-speaking volunteers in the Spanish American struggle for independence. See Alfred Hasbrouck, *Foreign Legionaries in the Liberation of Spanish*

South America (New York: Columbia University Press, 1928); A. Curtis Wilgus, "Some Activities of United States Citizens in the South American Wars of Independence, 1808–1824," *Louisiana Historical Quarterly,* vol. XIV (1931), pp. 182–203; and Charles Lyon Chandler, *Inter-American Acquaintances* (Sewanee, Tenn.: The University Press, 1917).

3. The Arcadian Academies which sprang up in Italy around 1700 had a strong echo in the Portuguese-speaking world. Members of such societies steeped themselves in the classics and spoke of the happy, natural life of shepherds and hunters removed from the vices of civilization. Although there was no formally organized academy in Minas Gerais, members of the intelligentsia adopted classical names, wrote eclogues in the style of Virgil, read the Encyclopedists and the American Declaration of Independence, and dreamed of improving the economic and political life of the Captaincy. See Oswald de Andrade, *A Arcadia e a Inconfidência Mineira* (São Paulo, 1945). In Samuel Putnam's *Marvelous Journey: A Survey of Four Centuries of Brazilian Writing* (New York, 1948), Chap. VII is entitled "Arcady and the Rights of Man."

4. Claudio Manuel da Costa (1729–1789) translated Adam Smith's *Wealth of Nations* and wrote an epic poem, *Vila Rica,* about the mining capital which was admired by the romantic poet Almeida Garret. It is thought that he composed the introductory *Epístola a Critilo* of the *Cartas chilenas.* The author of that satire, ridiculing the Governor of Minas Gerais and the conditions which led to the Inconfidência Mineira, was Tomás Antônio Gonzaga, known for *Marília de Dirceu,* a collection of Arcadian love poems written to a young girl with the imaginary name Marília. The work first published in 1792 went through numerous editions and combines descriptions of the region with moving passages of tenderness:

   Tu não verás, Marília, cem cativos
   Tirarem o cascalho e a rica terra,
   Ou dos cercos dos rios caudalosos,
      Ou da minada serra.

   Não verás separar ao hábil negro
   Do pesado esmeril a grossa areia,
   E ja brilharem os granetes de ouro
      No fundo da bateia.

   You will not see, Marília, one hundred slaves
   Draw pebbles and rich earth
   Either from the beds of mighty rivers
      Or from the mined mountains.

   You will not see the skillful Negro separate
   The coarse sand from the heavy gravel
   And nuggets of gold shining
      In the bottom of the pan.

   See Almir de Oliveira, *Gonzaga e a Inconfidência Mineira* (São Paulo,

1948); Joaquim Ribeiro, *As cartas chilenas e a Inconfidência Mineira* (Rio de Janeiro, 1950); Afonso Arinos de Melo Franco, *Critilo-Cartas chilenas* (Rio de Janeiro, 1940); and Caio de Melo Franco, *O inconfidente Cláudio Manuel da Costa* (Rio de Janeiro, 1931). The best editions of *Marília de Dirceu* are that of Alberto Faria (Rio de Janeiro, 1922) and the *Obras completas,* M. Rodrigo Lapa, ed. (São Paulo, 1942).

5. Maciel, who was deported to Angola after the trial, was allowed great freedom. Being a mineralogist, he examined the rock formations of the region and asked permission to develop the local iron mines. An opinion regarding the project was made by Manoel Ferreira da Câmara Bethencourt e Sá for the Crown and is quoted in Marcos Carneiro de Mendonça, *O Intendente Câmara,* 2d ed. (São Paulo, 1958), pp. 290–295: "Memória analítica sobre a memória escrita e enviada do degredado de Angola pelo inconfidente D. José Alvares Maciel, sobre a fábrica de ferro de Nova Oeiras . . . officio relativo à fábrica de ferro existente em Galingo, no Reino de Angola" Maciel was allowed to develop a foundry at Trombola in 1799. His death soon after was greatly lamented.

6. The *Excursão montanistica* was published a number of times and can readily be consulted in Alberto Sousa, *Os Andradas* (São Paulo, 1922), vol. III, pp. 507–530, where it is entitled *Viagem mineralogica na Provincia de S. Paulo por José Bonifacio . . .* .

7. The role of Minas Gerais was all important in the Independence Movement, and Câmara was a distinguished participant in every delegation connected with the national scene from 1820 to 1822. See Salomão de Vasconcelos, *O Fico: Minas e os mineiros na Independência* (São Paulo, 1937).

8. "Com a ciência, satisfez o que a razão cosmopolita devía à natureza. Com a acção, pagou o que a patria devía ao cidadão." There are many works on the Independence Movement and the role of the veteran scientist who crystallized the feelings of Brazilians at the time. The following are particularly useful: José Feliciano de Oliveira, *José Bonifacio e a indepêndencia* (São Paulo, 1955); Oliveira Lima, *O Movimiento da Independência 1821–1822* (São Paulo, 1922); Francisco Adolfo de Varnhagen, "História da independência (revista e anotada pelo Barão do Rio Branco com um índice onomástico)," tomo 173 of the *Revista do Instituto Histórico Geográficao Brasileiro* (1938); Tobias Monteiro, *História do Império: A elaboração da independência* (Rio de Janeiro, 1927). For a study of Masonry in the Independence Movement see Manuel Joaquim de Meneses, *Exposição histórica da Maçonaria no Brasil* (Rio de Janeiro, 1857).

9. The great desire of Spain's South American kingdoms for autonomy might have been satisfied, the author believes, without the bloody conflict which eventually took place, if a plan attributed to the Count of Aranda and presented to Charles III had been implemented. This politically far-sighted plan was based upon the consequences which would have been inevitable in the

kingdom as a result of Spain's recognition of the independence of the United States and consequently of the ideas embodied in the Declaration of Independence of 1776.

Aranda's plan proposed that Spain give up her American possessions and establish in their place three kingdoms, each under a Prince of the reigning Spanish dynasty. It is curious to observe that the first American insurgents in Mexico as well as in the southern countries linked their aspirations for independence to recognition of the sovereignty of Ferdinand VII. The Mexican Declaration of Independence proclaimed by Agustín de Iturbide in 1821, the Plan of Iguala, provided for a constitutional monarchy headed by Ferdinand VII or another Prince of the House of Bourbon and Austria as ruler of the new Mexican nation.

10. Salvador de Madariaga, *Bolívar* (New York, 1952), pp. 487–488.

11. Armando Alba in his preface to Cañete's *Guía histórica.* The life of Villalba is the subject of a book by Ricardo Levene, *Vida y escritos de Victoriano de Villalba* (Buenos Aires, 1946).

12. See José Joaquin Izquierdo, *La primera casa de las ciencias en México: El Real Seminario de Minería, 1792–1811* (Mexico City, 1960).

13. As a final footnote on the dramatic epic of the Conquest and settlement, a paragraph should be cited from the Introduction of Salvador de Madariaga's *El auge del imperio español en América* (Buenos Aires, 1955):

> The Spanish-American Empire, whose history begins in 1492 and ends during the first third of the nineteenth century, is an important element in the Hispanic pattern of the history of man. It can only be fitted into the general design if it is studied as such, as an historical entity with a beginning, a growth, an evolution and a disintegration which is not a mere final death, but a political death and an historical transfiguration. The fact that from Río Grande to Patagonia, people speak Spanish and look back on three centuries of Spanish life has a significance of its own which can be variously estimated, extravagantly exaggerated or contemptuously dismissed, but which in every case remains a fact, and a fact alive. The Hispanic body politic is no more; the Hispanic *body historic* lives on, no matter how absent-minded, divided against itself, unaware of its own existence, self-destructive even. It lives on.

Another discerning Spaniard, Don Luis Díez del Corral, in his thought-provoking *Del Nuevo al Viejo Mundo* (Madrid, 1963) saw how Spain, for all her intelligence and passion, did not know how to write the last chapter of emancipation. She did, however, write many of the most fortunate chapters in America's history, and as the Mexican educator and statesman Jaime Torres Bodet, former Director General of UNESCO, pointed out: "She wrote the most glorious chapters as well."

Chronological Table of Discovery and the Development of Mining in the New World

The chronological table combines several of the themes which run through this book. In addition to dates concerning mineral discoveries, it includes information on first-hand reports, chronicles, geographical charts, and explorations during the first 150 years after the discovery of the New World. Of particular interest are the items about the expeditions of Spaniards in the Pacific. They demonstrate that after discovering that ocean in 1513 and crossing it in 1520 in the Magellan–Elcano Expedition, the Spanish continued to explore in that region, sailing from El Callao, Peru, and the western ports of New Spain. Continuous communications were maintained between the Philippine Islands and the Viceroys of New Spain for the 250 years during which these high officials had administrative control over the islands. These items are listed not only to show their historical and literary importance, but also to refute the oft-repeated statement that the Spaniards concealed as much as possible the things that happened in the New World.

Prejudice against the Spanish accomplishments in America has reached such a pitch that it is even said that the Spaniards paid attention exclusively to the extraction of precious metals and neither observed nor wrote anything about the fauna, flora, or natural sciences. In a book, in other ways very accurate and of indubitable merit, Victor W. von Hagen's *South America Called Them: Explorations of the Great Naturalists La Condamine, Humboldt, Darwin, Spruce* (New York: Alfred A. Knopf, 1945), the author claims that the natural history of America begins on a certain day in 1734 in the Académie des Sciences in Paris when La Condamine was assigned the task of measuring a degree of latitude in Ecuador, high in the Andes, a few miles to the north of Quito. The gross inaccuracy of this assertion can be observed from the items included in this table.

1492 The discovery of America. Columbus makes his first voyage to the Bahamas, Cuba, and Santo Domingo (Hispaniola).

1493 Columbus's second voyage. He touches the Windward Islands, Puerto Rico, Santo Domingo, Jamaica, and Cuba. The first Spanish miners arrive with this expedition. Pope Alexander VI issues several Bulls granting to the Catholic Kings the properties of the lands and islands discovered by Columbus and making the meridian line to separate the lands which in the future will belong to Spain and Portugal.

1494 The Treaty of Tordesillas between Spain and Portugal modifies the Papal Bulls and the demarcation line established by them.

1498 Columbus's third voyage. On August 10 he discovers and disembarks on the South American mainland on the coast of Venezuela. Vicente Yáñez Pinzón sails along the coast of Brazil and explores the mouths of the Orinoco and Amazon.

1499 Alonso de Ojeda, leaving Cádiz in May with Amerigo Vespucci in his company, explores the mouth of the Rio Doce and the coast as far as Venezuela.

1500 On March 8, Pedro Alvarez Cabral at the head of a large fleet with more than 1,000 men leaves Lisbon and on April 23 lands near Porto Seguro, between Bahia and Rio de Janeiro.

1500–01 The first map of America is published by Juan de la Cosa.

1502 Columbus's fourth voyage. He discovers and explores the Caribbean coast of Central America and finds indications of gold. Fernão de Noronha obtains a contract from King Manuel to export brazilwood from Portuguese America.

1512 Ponce de León discovers Florida.

1513 Vasco Núñez de Balboa beholds the Pacific on September 24.

1516 Juan Dias de Solís discovers an estuary of the Río de la Plata.

1516–19 Cristóvão Jacques establishes a *feitoria* at Pernambuco and destroys French brazilwood trading posts.

1519 Hernán Cortés lands in Veracruz and occupies Mexico City (Tenochtitlán) on November 8. Alvarez de Pineda discovers the delta of the Mississippi. The *Suma de geografía* by Fernández de Enciso, the first geographical treatise on America, is printed in Seville.

1519–22 Magellan's ship *Vittoria,* arriving in Spain under the command of Sebastián Elcano after the Commander's death, makes the first circumnavigation of the globe from east to west.

1520 Cortés sends Diego de Ordaz to explore the isthmus of Tehuantepec and establish a means of communication between the Gulf of Mexico and the Pacific. Pedrarias de Avila founds the city of Panama. Lucas de Ayllon discovers the region of North Carolina ("Chicora") in North America.

1521 On August 13, Cortés retakes the city of Mexico, completing the conquest of "New Spain of the Ocean Sea."

1524 Tin and copper mines are operated in Tasco, Mexico, and the first smelting operations are started. Mines are discovered in Sultepec. The first of Hernán Cortés's *Cartas de relación al Emperador* is published in Zaragoza. Aleixo Garcia, a Portuguese who had accompanied Solís, explores the interior from the coast as far inland as present-day Paraguay and Bolivia and sends back samples of silver. Pedro de Alvarado founds the city of Santiago de los Caballeros (Guatemala). Lucas Vázquez de Ayllón in his second expedition explores the eastern coast of North America up to the Chesapeake Bay.

1525 Diego Alvarado founds the city of San Salvador. Esteban Gómez explores from Florida up to Labrador, sailing by the mouths of the Delaware, Hudson, and Connecticut Rivers. Operations begin in the first silver mines in New Spain at Morcillo, Jalisco, and Villa del Espíritu Santo in Compostela, Nayarit.

1525–26 García Jofre de Loaisa makes his expedition across the Pacific to the Moluccas. Sebastián Elcano is chief pilot and the young Andrés de Urdaneta his assistant. During the expedition, Loaisa and Elcano die, and of the original 400 men and 7 ships only

8 men, including Urdaneta, return to Spain in 1 ship. Sailing via the Cape of Good Hope, they complete the second trip around the world.

1526–28 Sebastian Cabot explores the Río de la Plata.

1527 On October 31 an expedition commanded by Alvaro de Saavedra Cerón, with 3 ships and 100 men furnished by Cortés at the Emperor's request, sails from Zihuatanejo, Zacatula, Mexico. Ordered to discover the fate of the Loaisa expedition, it touches the Ladrones and Philippine Islands.

1528–36 Alvar Núñez Cabeza de Vaca marches from Florida to Mazatlán on the Pacific coast (8 years and 6,000 miles), following the disaster of Pánfilo de Narváez's expedition.

1530–33 Martím Afonso de Sousa eliminates French trading activities on the Brazilian coast and founds São Vicente and Piratininga (São Paulo). The first sugar mills are erected in Brazil.

1531 Diego de Ordaz makes his way up the Orinoco from the Gulf of Paria, a point above its confluence with the Meta. Organized Portuguese *entradas* or *bandeiras* begin to explore the interior in search of gold and silver.

1531–33 Francisco Pizarro conquers Peru.

1533–34 Pedro de Heredia founds Cartagena de Indias and explores the Magdalena River.

1534 Silver mines are worked in Tasco, Mexico, for the first time, and Cortés himself opens the first tunnel, or adit, in the New World. Brazil is divided among twelve donataries in huge land grants similar to British proprietory colonies in North America. Juan de Ayolas explores the Río de la Plata.

1534–35 Sebastián de Belalcázar founds the cities of Quito and Guayaquil.

1535 The first mints are established in Mexico City, Lima, and Santa Fe (Colombia). Francisco Pizarro founds Ciudad de los Reyes (Lima). Cortés explores both coasts of the Gulf of California. The first edition of the first part of the *General y natural historia de las Indias* by Gonzalo Fernández de Oviedo, appointed

General Historian of the Indies in 1532 is published. Fray Tomás de Berlanga, Bishop of Castilla del Oro, discovers the Galápagos Islands. The Espíritu Santo mine in the Cerro de Jolotlán, District of Tepic (now Nayarit), is discovered, the first Mexican bonanza. The first printing press in America is established in Mexico City.

1535–37 Almagro's expedition travels from Cuzco in the valley of Maipú, Chile, along the Incas' Andean road.

1536 Belalcázar founds the cities of Popayán and Cali, Colombia; Pedro de Mendoza founds the city of Buenos Aires. Duarte Coelho, with wealth acquired in the Orient, settles Pernambuco, the most successful of the private Brazilian Captaincies.

1536–38 Gonzalo Jiménez de Quesada travels up the Magdalena River, reaches the savannas of Bogotá, and founds the city of Santa Fé.

1537–38 Juan de Ayolas and Domingo Martínez de Irala found the city of Asunción at the confluence of the Paraguay and Paraná Rivers.

1539–41 Hernando de Soto, having been with Pizarro in the conquest of Peru, explores the regions bordering the Mississippi and dies there.

1540 Alvar Núñez Cabeza de Vaca makes his second march, from Santa Catarina on the coast of Brazil to Asunción on the Paraná River, passing the Iguazú Falls. Francisco Vázquez de Coronado sets out to discover the legendary city of Cíbola and Quivira journeys from New Spain to present-day Kansas City. Coronado's captain, Don García López de Cárdenas, discovers the Grand Canyon of the Colorado River. The first mines are discovered in Zacatecas.

1541 Valdivia founds the city of Santiago de Chile.

1541–42 Francisco de Orellana travels down the Amazon from the Napo River in Ecuador to its mouth.

1542 The New Laws are promulgated "for the peservation of the natives [of the Indies], and their good government and the preservation of their persons." Gold deposits are opened in Carabaya, La Paz, Oruro, and other places in Upper Peru (modern

Bolivia) at an altitude of more than 12,000 feet. Juan Rodríguez Cabrillo explores the coasts of Upper California as far as the 37th parallel. On November 1 an expedition under the command of Ruy López de Villalobos sails from Navidad (Jalisco, Mexico) with 6 ships and 370 men; before reaching the Philippines, he discovers Revillagigedo Island, the Marshall archipelago, the Carolines, and the Palaus.

1545 The first silver mine is discovered in the Mountain of Potosí, Bolivia, at an altitude of more than 15,000 feet. Valdivia founds the city of La Serena (Chile).

1546–48 The chief silver lodes of the mines of Zacatecas, Guanajuato, and Santa Bárbara, Mexico, are discovered.

1549 Temascaltepec mines are discovered in the state of Mexico. Tomé de Sousa is appointed Governor General of Bahia, purchased by King João III the previous year from the Pereira Coutinho family. Bahia is made the principal Crown Captaincy. The power of donataries is restricted by royal decree. Father Manuel da Nóbrega with other Jesuits establishes Christianity among Indians following the King's wishes: "The principal reason motivating my decision to settle the land of Brazil is in order that the people of that land may be converted to our Holy Catholic Faith." Captain Alonso de Mendoza founds the city of Santa María de la Paz (Bolivia).

1550 Valdivia founds the cities of La Concepción and Valdivia, Chile, and rich gold mines are discovered in Confines and Quilacoya.

1551 The Universities of Mexico City and Lima are founded. The first mining works are begun in Real del Monte, Pachuca District, in the present state of Hidalgo. The first bishopric is established at Bahia.

1552 The first edition of Padre Bartolomé de Las Casas's *Destrucción de las Indias* is published in Spain. Ginés Vásquez de Mercado discovers the iron mines of Cerro de Mercado, Durango, Mexico. The first edition of Francisco López de Gomara's *On the Conquest of Mexico* is published.

1552–55 The famous silver mines of Pachuca, Fresnillo, Sombrerete, and Chalchihuites are discovered and exploitation begun in New Spain.

1553 The first edition of Pedro Cieza de León's *Crónica del Perú* is printed in Seville.

1554 Bartolomé de Medina in Pachuca, Mexico, invents and applies his "patio method" for processing silver by the use of mercury.

1555 The first edition of Alvar Núñez Cabeza de Vaca's *Naufragios y comentarios* is printed in Valladolid. It is translated into Italian in 1556 and into English in 1571. The French attempt to settle in the Bay of Guanabara (Rio de Janeiro).

1557 The first edition in Latin of the lectures delivered at the University of Salamanca by the Dominican Padre Francisco de Vitoria in 1537 and 1538, *De Indis* and *De Jure Belli,* is published; it is also published in Lyons in French. In these lectures the Crown of Spain's legal right to conquest and possession of the Indies is disputed. The first Spanish edition is published in Salamanca in 1565.

1558 The Portuguese Henrique Garcés discovers mercury in Peru.

1560 Pedro Menendez de Aviles founds the city of St. Augustine in Florida.

1561 The mining ordinances for the Viceroyalty of Peru are promulgated.

1564 The expedition of Miguel López de Legazpi sails from the port of Navidad, Mexico, to conquer and Christianize the Philippine Islands, which henceforward become part of the realm of the Viceroyalty of Mexico. The cosmographer of the expedition is Padre Andrés de Urdaneta, an Augustinian who made the second voyage around the world as a member of the Loaisa expedition in 1526.

1565 The flagship of the López de Legazpi expedition, commanded by Felipe de Salcedo and piloted by Urdaneta, enters Acapulco, returning from the Philippines on October 8. After the unsuccessful attempts of Bernardo de la Torre in 1544 and Ortiz de Retes in 1545, this expedition succeeds in finding the return route from the Philippines to New Spain. The Manila Galleon will use this route for the next 250 years.

1566 Mercury mines, of great importance for extracting silver from its ores, are discovered in Huamanga and Huancavelica, Peru.

1567 Diego Losada founds the city of Caracas. Governor Mem de Sá from Bahia defeats the French and refounds Rio de Janeiro.

1567–68 The first expedition by Alvaro Mendaña de Neyra from Callao discovers the Solomon Islands, among them Guadalcanal and Isabel, and returns to Acapulco.

1570 Francisco Hernández, appointed Physician-in-Chief of the Indies by the King, sails for Mexico.

1570–76 Pero de Magalháes de Gândavo's *Tratado da terra do Brasil* and *História da Provincia de Santa Cruz,* written partly to encourage settlement, is published. The latter contains an introduction in verse by the epic poet Luis de Camões.

1571 Miguel López de Legazpi founds the city of Manila.

1571–72 Pedro Fernández de Velasco perfects the adaptation of Bartolomé de Medina's mercury amalgamation process to the conditions of the Peruvian-Bolivian high plateau.

1572 The famous mint is built in Potosí, Bolivia.

1574 Mining begins at Charcas in the present state of San Luis Potosí.

1578 The Viceroy Don Francisco de Toledo draws up new mining ordinances for Peru.

1580 Philip II of Spain becomes Philip I of Portugal, uniting the Peninsular Crowns.

1584 Philip II promulgates the *Ordenanzas del Nuevo Cuaderno,* a summary of mining laws and regulations for America.

1587 Gabriel Soares de Sousa draws up his *Tratado descritivo do Brasil em 1587,* the most important sixteenth-century chronicle about Portuguese America. It describes the land, sugar plantations, fauna, and flora as well as the gold, silver, and precious stones to be found in the interior.

1589 The first edition of part of José de Acosta's *Natural and Moral History of the Indies* is published in Salamanca, in Latin, following Acosta's return from a sojourn of sixteen years in America, where he gathered a very large amount of scientific information on geography, natural science, and ethnography. The first complete Spanish edition is printed in Seville in 1590, and by 1598 it has been translated into Italian, French, English, German, and Flemish.

1595 Mendaña, on his second expedition to the Pacific from Callao, discovers the Marquesas and Santa Cruz Islands.

1596 Diego de Montemayor founds the city of Monterrey, Mexico.

1596–02 In the course of two voyages, Sebastián Vizcaíno explores and makes an accurate chart of the coast of California from the port of Acapulco to the 42nd parallel.

1598–05 Following the discovery of small alluvial deposits of gold in the south of Brazil, Governor General Francisco de Sousa moves his residency from Bahia to the Captaincy of São Vicente and receives the title of Superintendant of Mines.

1599 Natal is founded on the northeast corner of Brazil.

1603 A new code of law, the *Ordenações Filippinas,* set up by Philip III of Spain (Philip II of Portugal) to govern overseas Portuguese possessions is established. The code remains basically in effect until 1822. A *Regimento das minas* is promulgated to handle mining activities.

1604–14 The *Conselho da India* (Council of India), similar to Spain's *Consejo de las Indias,* is created to implement the new laws and coordinate colonial questions in Portuguese America and Asia.

1606 A mint is established in Zacatecas.

1606–07 Pedro Fernández de Quirós discovers the archipelago of the New Hebrides. During this expedition the sailor Luis Vaez de Torres explores the southern coast of New Guinea and discovers the strait which still bears his name. On his return he makes port in Mexico.

1609 Alvaro Alonso Barba invents the "kettle and cooking" process for silver-bearing ores, in Tarabuco, Bolivia. The Inca Garcilaso de la Vega's *Comentarios reales* (a history of the Inca Empire) is published. This publication is followed in 1617 by his *Historia general del Peru.*

1614 The University of Córdoba is founded in Argentina. Brazilians and local Indians commanded by Jeronymo de Albuquerque and Alexandre de Moura end French attempts to settle Maranhão.

1624 The University of Chuquisaca is founded in Bolivia.

1624–25 Bahia, the capital of Brazil, is captured by the Dutch and then lost to the victorious fleet of D. Fadrique de Toledo Osorio. Lope de Vega writes *El Brasil restituído.*

1629 Small gold deposits are discovered at Paranaguá in southern Brazil. Additional small finds are made in 1648 and 1658.

1630–54 The Dutch West India Company occupies northeastern Brazil with headquarters at Recife, the capital of Pernambuco.

1632 The first edition is published of *La Verdadera historia de la Conquista de México* by Bernal Díaz del Castillo, a participant in the Conquest of New Spain, written in Guatemala when he was over eighty years of age. Mines are opened at Batopilas, Chihuahua.

1637–39 The Portuguese Pedro Teixeira explores the region of the Yupur River, a tributary of the Amazon. He discovers gold and takes possession in the name of the Portuguese Crown.

1640 The first edition of *El Arte de los metales,* by Alvaro Alonso Barba, is published in Madrid; it is to be issued more than thirty-five times, in six languages. Portugal declares its independence from Spain. The eldest son of the House of Braganza is named Prince of Brazil.

1642 The *Conselho Ultramarino* is established as the supreme overseas body to coordinate Portuguese colonial policy.

1643 Salvador de Sá is appointed Administrator of Mines in São Paulo and receives *Regimento* for special privileges in regard to discoveries.

1648–51 Antônio Rapôso Tavares makes one of the most remarkable bandeiras ever to explore the interior, traveling from São Paulo to Paraguay and the Chaco, west to the silver region of the Andes, and then northeast down the Amazon to Belem do Pará, a trek of over 6,000 miles.

1654 The Dutch capitulate at Taborda and depart from Brazil.

1655 Padre Antônio Vieira obtains permission from King João IV for the Amer-Indians to live freely under native chiefs and receive spiritual guidance from the Jesuits.

1674–81 The expedition of Fernão Dias Pais, one of the longest and most celebrated of the bandeiras, explores the interior from the headwaters of the Rio das Velhas to the Sêrro do Frio zone, later to yield rich gold and diamond deposits.

1680 Settlement by the Portuguese of Colônia do Sacramento opposite Buenos Aires takes place to round out the Brazilian border to the Río de la Plata.

1684 The first edition of *History of the Conquest of Mexico* by Antonio de Solís is published.

1693–95 Large quantities of gold in the Rio das Velhas, Rio das Mortes, and Rio Doce, north of Rio de Janeiro and west of Vitoria, are discovered. An influx of prospectors begins into the region that will come to be called Minas Gerais.

1695 The bandeirantes from São Paulo discover rich gold placers of the Rio das Velhas, Minas Gerais, the site of an extraordinary bonanza for almost a century.

1701 Governor Artur de Sá in Rio de Janeiro promulgates a new mining code for Brazil which is accepted and enacted by the Crown in 1702.

1708–09 A war is fought between the *Emboabas* (outsiders from the seacoast and Europe) and the Paulistas, the initial discoverers of the gold streams.

1709 São Paulo, including the Minas de Ouro, is promoted to a Captaincy.

1711 The Jesuit André João Antonil publishes in Lisbon *Cultura e opulência do Brasil por suas drogas e minas,* an excellent account of agriculture and mining.

1719 Gold is discovered in Cuiabá, Mato Grosso.

1720 From this date all Governors General of Brazil receive the additional title of Viceroy.

1724–25 Gold is found at Araguaya in Goiás, Brazil.

1725 The mint at Villa Rica de Ouro Prêto opens.

1726–28 Diamonds are discovered in the Sêrro do Frio region of Minas Gerais.

1727 Coffee is introduced into Brazil.

1730 Sebastião da Rocha Pinto publishes *História da América Portuguesa* in Lisbon.

1735–44 The expedition of C. M. de La Condamine, of the Academy of Sciences in Paris, sets out to measure a degree of the meridian in Ecuador.

1736 Antonio de Ulloa, who accompanied the La Condamine expedition as a representative of the Spanish government, discovers platinum in the mines of Chocó, Colombia, and describes it in 1748 in his *Relación histórica del viaje a la América meridional,* written in collaboration with Don Jorge Juan.

1740 The contract system for collecting diamonds is adopted by the Portuguese Crown.

1744 Goiás in Brazil is made a Captaincy.

1748 Mato Grosso is elevated to a Captaincy.

1748–57 Don José de la Borda, a mining entrepreneur of Tasco, has an extraordinary bonanza in his mine, La Lajuela.

1750 The treaty of Madrid between Portugal and Spain establishes principle of *uti possidetis* in determining boundaries. Pombal begins his long career as Portugal's chief Minister.

1760 Antonio Obregón y Alcocer (later Count of La Valenciana) begins work in the mother lode in Guanajuato, Mexico, as he exploits the La Valenciana mine.

1761 *Comentarios a las Ordenanzas de Minas* by the Mexican jurist Don Francisco Xavier de Gamboa is published in Madrid. The Treaty of Pardo annuls the Treaty of Madrid.

1762 A great bonanza is discovered at Real del Monte; Romero de Torres extracts some 31 million dollars in silver.

1763 The capital of Brazil is transferred from Bahia to Rio de Janeiro.

1767–84 Fray Junípero Serra walks from Mexico to Upper California and founds the missions of San Diego (1769), San Carlos de Monterrey (1770), San Antonio de Padua and San Gabriel (1771), San Luis Obispo (1772), and San Francisco and San Juan Capistrano (1782). He dies in 1784 and is buried in Monterrey in Mexico.

1771 Pombal abolishes the diamond contract system in favor of sole management by the Crown, the *Real Extracção.*

1773 The great silver mines are discovered at Catorce in the state of San Luis Potosí.

1774 Juan Pérez, in the corvette *Santiago,* sails from Mexico and explores the coast of California as far north as the 60th parallel.

1775 Francisco de la Bodega y Quadra and Bruno Hecela, sailing from the port of San Blas, Mexico, explore the northwest coast of America, covering the 17th to the 58th parallel.

1777 The Treaty of San Ildefonso between Portuguese- and Spanish-speaking America establishes borders which, with some modification, are similar to those of today.

1779 The Academy and Theoretical-Practical School of Metallurgy is inaugurated in Potosí, Bolivia.

1783 Charles III approves new mining ordinances for the Viceroyalty of New Spain which also provide for the creation of a college of mining in Mexico.

1788–89 Brilliant satirical verses of *Cartas Chilenas* by Judge

Tomás Antônio Gonzaga expose corrupt policies of the Portuguese Governor of Minas Gerais, Luis da Cunha Menezes.

1789 The *Inconfidência Mineira* is led by an army lieutenant and engineer, Joaquim José da Silva Xavier (Tiradentes).

1792 The Royal Seminary of Mining is inaugurated in Mexico City. Its first director is Fausto de Elhuyar, discoverer of tungsten. Conspirators in Minas Gerais are deported and Tiradentes is executed.

1794 Don Andrés Manuel del Río arrives in Mexico as professor of the Royal Seminary of Mining. In 1801, he discovers vanadium in the dark lead ores of Zimapán. *Ordenzas de las minas del Perú y demás provincias del Río de la Plata* drawn up by Don Pedro Vicente Cañete is published.

1803 The *Alvará,* or Decree, of 1803, is drawn up to replace Pombal's *Regimento Diamantino* of 1771.

1808 King João VI of Portugal comes to Brazil, fleeing the Napoleonic invasion. Napoleon's troops invade the Iberian Peninsula and Ferdinand VII is put under house arrest at Bayonne, France. Dom João, Prince Regent of Portugal, and 12,000 followers and members of the Court flee Lisbon in British vessels and take up residence in Rio de Janeiro. The ports of Brazil are opened to world trade.

1810 The Insurgent Movement begins in almost all the kingdoms and provinces of Spanish America, demanding political independence but reaffirming loyalty to Ferdinand VII: Venezuela (August 2), Argentina (May 21), New Granada (Colombia) (July 20), Paraguay (July 24), Ecuador (August 2), Mexico (September 15), and Chile (September 18). Many Spanish-American mining engineers join the Insurgent Movement and perish in the conflict.

1815 Brazil becomes a kingdom on equal status with Portugal.

1816 Río de la Plata, Argentina, declares its independence (July 9).

1818 Bolívar creates the republic of Gran Colombia in Angostura, Venezuela. San Martín's victories at Chacabuco and Maipú assure Chile's independence (April 5).

1821 Iturbide declares the independence of Mexico at Iguala, in the present state of Guerrero (February 7). San Martín declares Peru independent (July 6).

1822 The independence of Brazil is proclaimed, and Pedro I, son of João VI, is named Emperor (September 7).

1824 The Battle of Ayacucho (December 10) confirms the independence of Peru and all the other Spanish territories on the American continent.

Bibliography

The following bibliography is a broad-based one in the spirit of this book. It offers the reader a wide range of items from specialized technical studies and historical esoterica to general studies for the student to popular works for the armchair adventurer. Two objectives have been pursued in compiling this bibliography: first, to offer a guide to further reading and research to both the undergraduate student and the serious researcher; second, to offer a selection which would include not only works that concern the history of mining in colonial Hispanic America, but also some general studies of Latin American history that indicate the influence of mining and Spain's and Portugal's roles in the institutional life and culture of their New World colonies. In any event, it is far from exhaustive in either its general selections or its narrower, more technical citations. A complete coverage is far beyond the scope and size of this study.

This listing is divided into several sections by the type of references and by the subject matter. In most cases, the titles speak for themselves. Notes are appended to certain titles which might not appear to be germane to this study or which are of especial interest.

General dictionaries and encyclopedias are not included, but naturally a great deal of useful information on specific topics and individuals is available in such publications as *Diccionario enciclopedico Hispano-Americano de literatura, ciencia y artes,* 23 vols. and 3 appendices (Barcelona: Montaner y Simón, Editores, 1887–1910); *Diccionario geográfico, histórico y biográfico de los Estados Unidos Mexicanos,* Antonio García Cubas, ed., 5 vols. (Mexico, 1896); *Grande enciclopédia Portuguesa e Brasileira,* 37 vols. and 3 appendices (Lisbon and Rio de Janeiro: Editorial Enciclopédia, Limitada, 1935–1945); Manuel de Mendiburu, *Diccionario histórico-biográfico del Peru,* 8 vols, (Lima: Imprenta de J. F. Solis, 1874–1890), reprinted in 11 vols. [Lima: Imprenta "Enrique Palacios," 1931–1934 (1935)]; and Ernesto Shäfer, *Indice a la colección de documentos inéditos de Indias,* 2 vols. (Madrid, 1946–1947).

A great many document collections also are available, ranging from general collections such as *Gobernantes del Perú: Cartas y papeles, siglo XVI; documentos del Archivo de Indias,* Roberto Levillier, ed., 11 vols. (Madrid: Colección de Publicaciones Historicas de la Biblioteca del Congreso Argentino, 1921–1925) to more specialized collections such as *Colección de documentos para la historia de San Luis Potosí,* Primo Feliciano Velásquez, ed., 4 vols. (San Luis Potosí, Mexico, 1897–1899). It is worth noting that since any study of Potosí is bound up with the history of mining, general books, articles of bibliographical interest, and documents dealing with the Imperial City are worth consulting. In general, however, in document collections the references to mining usually are found in a matrix of miscellaneous information. These sources, therefore, would best be used by more serious students of the subject.

The same is true of the many chronicles of Conquest, contemporary histories, and geographical descriptions that were published by conquistadors, colonists, and travelers. All of them convey a sense of the quality of colonial life and therefore ought to be read by students of colonial Hispanic America. However, with a few exceptions, the amount of material they contain specifically on mining would reward only the attentions of a more serious researcher. A few of the more important works of this type are José de Acosta, S.J., *Historia natural y moral de las Indias en que se tratan las cosas notables del cielo, y elementos, metales, plantas, y animales dellas; y los ritos, y ceremonios, Reyes, y govierno, y guerras de los Indios* (Seville, 1590); Antonio de Alcedo, *Diccionario geográfico-histórico de las Indias Occidentales ó América; es a saber: De los reynos del Perú, Nueva España, Tierra Firme, Chile, y Nuevo reyno de Granada,* 5 vols. (Madrid, 1786–1789); Pedro Vincente Cañete y Dominguez, *Guía histórica, geográfica, física, política, civil y legal del Gobierno e Intendencia de la provincia de Potosí* (1787); Alonso de Carrió de la Vandera (pseud. Concolorcorvo), *El lazarillo de ciegos caminantes* (Gijon, Spain, 1773); Antonio de Herrera y Tordesillas, *Historia general de los hechos de los castellanos en las islas y tierra firme del mar oceanico,* 3 vols. (Madrid, 1601–1615).

While this study was in preparation a series of volumes was published in connection with the Sixth International Mining Congress held in Madrid in June 1970 under the title *La Minería Hispana e Iberoamericana: Contribución a su investigación historica* [León, Spain: Catedra de San Isidoro (Apartado 126), 1970–1971]. The most important is vol. I, *Ponencias del Coloquio Internacional sobre Historia de la Minería,* which devotes 420 pages to mining in the New World. Among the articles of particular interest are Paulino Castañeda Delgado, "El tema de las minas en la ética colonial española"; Demetrio Ramos, "Ordenación de la minería en Hispanoamérica durante la época provincial (siglos XVI, XVII y XVIII)"; S. Martínez

Constanzo, "La minería rioplatense en el último tercio del siglo XVIII"; Guillermo Lohmann Villena, "La minería en el marco del Virreinato peruano"; and José Tudela de la Orden, "La minería y la metalurgia de la América española en los manuscritos de las bibliotecas de España." Volumes II and III are a facsimile reprint of Eugénió Maffei and Ramón Rúa Figueroa's *Apuntes para una biblioteca Española . . . relativos a . . . las riquezas minerales* (1871–1872) which is listed in the Bibliography. Volume VII is a critical edition of *Capitulos sobre los metales de las etimologias de Isidoro de Sevilla,* Manuel C. Díaz y Díaz, trans. and ed. Five volumes are bibliography and archival guides: vol. IV, *Apuntes para una bibliografia minera Española e Iberoamericana (1870–1969);* vol. V. *Archivo general de Simancas: Indice de documentación sobre minas (1316–1832);* vol. VI, *Documentos existentes en el Archivo de Indias—Sección de Guatemala;* vol. VIII, *Documentos existentes en el Archivo de Indias—Sección de Lima;* vol. IX, *Documentos existentes en el Archivo de Indias—Sección de México*. Volume X contains the index to the series.

Bibliographies and Guides, Dictionaries, and Statistical Collections

Aguilar y Santillán, Rafael, *Bibliografía geológica y minera de la República Mexicana,* Instituto Geológico, *Boletín,* no. 10 (Mexico: Oficina Tipográfica de la Secretaría de Fomento, 1898).

———, *Bibliografía geológica y minera de la Republica Mexicana: Completada hasta el año de 1904* (Mexico: Imprenta de la Secretaría de Fomento, 1908).

———, Departamento de Minas, *Bibliografía geológica y minera de la República Mexicana, 1905–1918* (Mexico: Secretaría de Hacienda, 1918).

———, Secretaría de la Economia Nacional, *Bibliografía geológica y minera de la República Mexicana* (Mexico: Talleres Gráficos de la Nación, 1936). Most of the material listed concerns the contemporary period, but there are listings of studies both of a historical nature and of interest as well as more modern descriptions of the geology of old districts.

Arnade, Charles W., "Una bibliografía selecta de la guerra de la emancipación en el Alto Perú," *Boletín de la Sociedad Geográfica y de Historia Potosí,* vol. XL (1953), pp. 159–169. Items on the historic mining region of Upper Peru during the Wars of Independence.

Bradley, Walter W., "Bibliography on Quicksilver," in *Quicksilver Resources of California,* California State Mining Bureau, *Bulletin* no. 78 (Sacramento, Calif., 1918). A few Latin American items. (See also Whitaker, *The Huancavelica Mercury Mine,* below).

Bureau for Economic Research in Latin America, Harvard University, *The Economic Literature of Latin America,* 2 vols. (Cambridge, Mass.: Harvard University Press, 1935–1936). Mainly lists books and articles published during the national period, but many items have a historical interest, particularly for economic history. Mining is one of the fields covered within each national listing.

Busto, Emiliano, *Estadística de la República Mexicana,* 3 vols. (Mexico, 1880). Volume II is devoted to mining; and contains figures on the colonial period.

Calderón, Abdón *Diccionario y vocabulario minero* (La Paz, 1939). Explanations of technical terms and jargon.

Crane, Walter R., *Index of Mining Engineering Literature, Comprising an Index of Mining, Metallurgical, Civil, Mechanical, Electrical, and Chemical Engineering Subjects as Related to Mining Engineering,* 2 vols. (New York: J. Wiley & Sons, 1909–1912). Has several items of historical interest.

Gumucio, Gonzalo, "Las mil y una historias de la Villa Imperial de Potosí," *La Razón* (La Paz, Dec. 17, 1950, and Jan. 7, 1951).

Halse, Edward, *A Dictionary of Spanish, Spanish-American, Portuguese and Portuguese-American Mining, Metallurgical and Allied Terms* (London: C. Griffin, 1926).

Hanke, Lewis, *The Sources Used by Bartolomé Arzáns de Orsúa y Vela for His History of Potosí: Jahrbuch für Geschichte von Staat, Wirtschaft und Gesellschaft Lateinamerikas . . .* Sonderdruck aus Band II (Cologne: Böhlau, 1965). Indispensable for the study of the early history of Upper Peru.

Lexis, Wilhelm, "Beiträge zur Statistik der Edelmetalle," *Jahrbücher für Nationalökonomie und Statistik,* vol. XXXIV (1879), 361–417. A classic study of mining and mineral statistics; see Soetbeer below.

Maffei, Eugenio, and R. Rúa Figueroa, *Apuntes para una biblioteca española de libros, folletos, y artículos, impresos y manuscritos, relativos al conocimiento y explotación de las riquezas minerales y á las ciencias auxiliares,* 2 vols. (Madrid: J. M. Lapuente, 1871–1872). Indispensable. An excellent starting point for the serious student.

Merrill, Charles W., et al., *Summarized Data of Silver Production.* U.S. Bureau of Mines, Economic Paper 8, 1930. An excellent summary of the statistics available on a worldwide basis.

Nieto, Felix, *Apuntes en forma de diccionario por el minero práctico* (Zacatecas, Mexico, 1891).

Orozco y Berra, Manuel, *Apuntes para la historia de la geografía en México* (Mexico, 1881). A great deal of interesting information on mining.

Pinelo, Antonio Leon, *Epitomé de la Biblioteca Oriental i Occidental Naútica i Geográfica* (Madrid, Juan González, 1629). This first bibliography in which there are numerous useful references to mining in the New World was reprinted with many additions by Andres Gonzalez de Barcia (Madrid, Francisco Martinez Abad, 1737–1738).

Sellerier, Carlos, *Data Referring to Mexican Mining* (Mexico: F. P. Hoeck & Co., 1901). A handy collection of facts and figures.

Soetbeer, Adolf, *Edelmetall-produktion und Werthverhältniss zwischen Gold und Silber seit der Entdeckung Amerikas bis zur Gegenwart* (Gotha, 1879). The classic study of world production of gold and silver. Some of the estimates are subject to question, particularly in the light of more recent research, but as a whole it is basically sound.

Vásquez Machicado, José, *Catálogo de documentos referentes a Potosí en el Archivo de Indias en Sevilla,* Armando Alba, Preface; Humberto Vázquez Machicado, Prologue (Potosí, 1964).

Zubiria y Campa, Luis, "Bibliografía minera, geológica y mineralógica del estado de Durango," *Memorias de la Sociedad Científica "Antonio Alzete,"* vol. 38 (Mexico, 1919), pp. 177–198.

General Histories

There are a number of general histories not dealing primarily with mining which deserve attention. Some deal with countries, others with districts or specific periods in which mining was of great importance.

Abreu, João Capistrano de, *Capítulos de história colonial (1500–1800)* (Rio de Janeiro: M. Orosco & C., Impressores, 1907). A recent edition annotated and with a preface by José Honório Rodrigues was published in 1954 by the Sociedade Capistrano de Abreu in Rio de Janeiro by Libraria Briguiet.

Alamán, Lucas, *Historia de Mexico,* 5 vols. (Mexico, 1849–1852).

Altamira y Crevea, Rafael, *España en América* (Madrid, 1898).

Bancroft, Hubert Howe, *History of Mexico,* 6 vols. (San Francisco, Calif.: The History Company, Publishers, 1886–1888). This history comprises vols. IX to XIV of *Bancroft's Works.*

Calmon, Pedro, *História do Brasil,* 2d ed., 7 vols. (Rio de Janeiro, 1963).

Haring, Clarence Henry, *The Spanish Empire in America* (New York: Oxford University Press, 1947). Contains an important discussion of colonial mining.

Holanda, Sergio Buarque de, *Historia geral da civilização Brasileira,* tomo I, vols. 1 and 2: *A época colonial;* tomo II, vol. I: *O Brasil Monarquico* (São Paulo, 1960–1963).

Madariaga, Salvador de, *The Rise of the Spanish American Empire* (New York: The Macmillan Co., 1947). A Spanish edition appeared as *El auge del imperio español en América* and was published by Editorial Sudamericana in Buenos Aires in 1955.

______, *The Fall of the Spanish American Empire* (New York: The Macmillan Co., 1947). A Spanish edition appeared as *El Ocaso del imperio español en América* and was published by Editorial Sudamericana in Buenos Aires in 1955.

Riva, Palacio, *México á través de los siglos,* 5 vols. (Barcelona, 1888–1889).

Rodrigues, José Honório, *Brasil: Período colonial,* Instituto Panamericano de Geografía e Historia, Comisión de Historia, Publicación no. 53 (Mexico, 1953).

Southey, Robert, *History of Brazil,* 3 vols. (London, 1810–1819).

Varnhagen, Francisco Adolpho de, *História geral do Brasil antes da sua separação e indepêndencia de Portugal,* 2 vols. (Madrid and Rio de Janeiro, 1854–1857), and 5th ed. (São Paulo, 1956).

Zavala, Silvio Arturo, *The Colonial Period in the History of the New World.* Abridgement in English by Max Savelle, publ. 239 (Mexico: Pan American Institute of Geography and History, 1962). A fuller Spanish edition entitled *El Mundo americano en la época colonial* was printed by Editorial Porrúa in its Biblioteca Porrúa, nos. 39 and 40, in Mexico City in 1967.

Contemporary Descriptions of Mines

This section of the bibliography is complemented by the one immediately following. They give lengthy and detailed descriptions of mines and their workings, sometimes with a survey of or comments on the social and economic life of the region.

Adams, William, *Actual State of the Mexican Mines* (London, 1825).

Alamán, Lucas, "Causas de la decadencia de la minería de Nueva España," United Mexican Mining Association, *Report of the Court of Directors,* Richard Heathfield, trans. (London, 1827).

Andrada e Silva, Jcsé Bonifacio, *Obras cientificas, politicas e sociais de José Bonifacio de Andrada e Silva,* Edgard de Cerqueira Falcão, ed. (São Paulo, 1965). Includes French and English versions of Bonifacio's works on mining.

Antonil, André Antônio (João Antônio Andreoni), *Cultura e opulência do Brasil por suas drogas e minas* (Lisbon, 1711). There is a reprint edited by A. P. Canabrava, Coleção Roteiro do Brasil, (São Paulo: Companhia Editora Nacional, 1967), vol. II. A reprint of the 1711 edition with a translation in French on facing pages was done by Andrée Mansuy (Paris: Institut des Hautes Etudes de l'Amérique Latine, Université de Paris, 1968). The classic study of Brazil in the early eighteenth century.

Arzáns de Orsúa y Vela, Bartolomé, *Historia de la Villa Imperial de Potosí,* Lewis Hanke and Gunnar Mendoza L., eds., 3 vols. (Providence, R.I.: Brown University Press, 1965). A discussion of the writing and significance of this study was presented by

Dr. Hanke in the Colver Lectures in 1965, printed with the title *Bartolomé Arzáns de Orsúa y Vela's History of Potosí,* (Providence, R.I.: Brown University Press, 1965).

Ballivián y Roxas, Vicente, *Anales de la Villa Imperial de Potosí* (Paris, 1872).

Capoche, Luis, *Relación general de la Villa Imperial de Potosí* (1585), with a preliminary study by Lewis Hanke, ed., Biblioteca de Autores Españoles, no. 122 (Madrid: Ediciones Atlas, 1959). A discussion of the writing and significance of this study was done by Dr. Hanke, *Luis Capoche y la historia de Potosí, 1545–1585,* no. 28 (Lima: Publicaciones del Instituto Riva-Agüero, 1959).

Castelazo, José Rodrigo de, *De la riqueza de las minas . . . de Real del Monte* (Mexico, 1820).

Eguía, José Joaquin de, *Memoria sobre la utilidad e influjo de la minería en el Reino, necesidad de su fomento y arbitrios de verificado, presentada al importante cuerpo de la minería* (Mexico, 1819).

Elhuyar, Fausto de, *Memoria sobre el influjo de la minería en la agricultura, industria, población y civilización de la Nueva España en sus diferentes epocas, con varios disertaciones relativos a puntos de economía pública conexos con el propio ramo* (Madrid: Imprenta de Amarita, 1825).

Fernández de Santillán, Felipe, *Relación del estado de algunas cosas de la Villa Imperial de Potosí y cerro rico de ella,* in *Colección de documentos inéditos para la historia de España,* vol. LII, pp. 451–453.

Hernández, James, *A Philosophical and Practical Essay on the Gold and Silver Mines of Mexico and Peru; Containing the Nature of the Ore, and the Manner of Working the Mines; the Qualities and use of Quicksilver; the Cleansing and Refining of These Metals . . . Translated from a Letter Wrote [sic] in Spanish, by Father James Hernandez of the Society of Jesus, Employed by His Catholic Majesty to Write *the Natural History of the West Indies* (London: Printed for J. Scott, 1755).

James, Edward, *Remarks on the Mines, Management, Ores, . . . of the District of Guanajuato* (London, 1827).

Kruse, Hans, ed., *Deutsche Briefe aus Mexiko, Mit Einer Geschichte des Deutsch-Amerikanischen Bergwerksverein, 1824–1838: Ein Beitrag Zur Geschichte des Deutschtums im Auslande* (Essen, Germany: G. D. Baedeker, 1923).

Martínez y Vela, Bartolomé, *Anales de la Villa Imperial de Potosí,* Publicaciones del Ministerio de Educación, no. 1, (La Paz: Biblioteca Boliviana, 1939).

Moglia, Raúl, ed., "Representación escéncia en Potosí en 1663," *Revista de Filología Hispanica* (Buenos Aires–New York), vol. V, no. 2 (April–June 1943).

Ontdeckinghe van Rijcke Mijnen in Brasil (Amsterdam: Gherdruckt voor I. van Hilten, 1639).

Orsúa y Vela, Bartolomé, see Arzáns de Orsúa y Vela.

Paso y Troncoso, Francisco del, ed., *El mineral de catorce: Noticias de minas* (Mexico: Editor Vargas Rea, 1956).

———, *Descripción de las minas de Pachuca* (Mexico: Editor Vargas Rea, 1951).

Pino Manrique, Juan del, "Descripción de la villa de Potosí" (1787), in Pedro de Angelis, ed. *Colección de obras y documentos relativos a la historia antigua y moderna de las provincias del Río de la Plata,* vol. I. (Buenos Aires, 1836).

Río, Andrés del, "Discurso sobre la ferrería de coalcomán," *Diario de México,* suppl. Mar. 18, 1810.

Sandoval y Guzmán, Sebastián, *Pretensiones de la Villa Imperial de Potosí* (Madrid, 1634).

Sola y Fuente, Gerónimo de, *Nuevo assiento . . . con el Gremio de mineros . . . de Guancavelica* (Lima, 1745).

Unánue, José Hipólito (?), "Historia de la mina de Huancavelica," *Mercurio Peruano* (Jan. 30, 1791), pp. 65–68.

Vives y Echeverría, Juan, *Derrotero de la Real Mina de azogues de Santa Bárbara en la provincia de Huancavelica en el Perú* (Cadiz, 1812); reprinted in Arana, *Minas de azogue,* pp. 49–76.

Reports Of Travelers and Scientific Investigators

Traveler and *scientific investigator* were often interchangeable terms in the eighteenth century. A great deal of invaluable information including minute descriptions, interviews, and even documents were included in the writings then. For this reason, they are generally excellent sources of contemporary information, often with the added fillip of intelligent and insightful commentary.

Acarete de Biscay, Mons., *An Account of a Voyage up the River de la Plata and Thence over Land to Peru* (London: Printed for Samuel Buckley, 1698).

Bougainville, Louis Antoine de, *Voyage autour du monde, par le frégate du roi la Boudese, et la flûte l'Etoile, En 1766, 1767, 1768 & 1769,* 2 vols., 2d ed., augum. (Paris: Chez Saillant & Nyon, 1772). This book was translated into English under the title *A Voyage Round the World Performed by Order of His Most Christian Majesty, in the Years 1766, 1767, 1768 and 1769,* translated by John Reinhold Forster and printed in London by J. Nourse in 1772. A Spanish edition translated by Josefina Gallego de Dantín appeared under the title *Viaje alrededor del mundo por la frigata del rey la "Boudese" y la fusta la "Estrella" en 1767, 1768 y 1769.* It was published in Madrid in two volumes by Calpe in 1921.

Courte de la Blanchardière, M. L'Abbé, *Nouveau voyage fait au Pérou auquel on a joint une description des anciennes mines d'Espagne, traduiste de l'Espagnol d'Alonso–Carillo–*

Lazo (Paris: De l'Imprimerie de Delagnette, 1751).

______, *A voyage to Pérou in the years 1745, 1746, 1747, 1748 and 1749* (London, 1752).

Eschwege, Wilhelm Ludwig, von, *Journal von Brasilien: Oder Vermischte Nachrichten aus Brasilien, auf wissenshaftlichen Reisen gesammelt von M. C. von Eschwege,* F. J. Bertuch, ed. 2 vols. in 1 (Weimar: Landes-Industrie-Comptoir, 1818). There is a translation into Portuguese of this volume by Lucia Furquim Lahmeyer entitled *Diario de uma viagem do Rio de Janeiro a Villa Rica, na capitânia de Minas Gerais no anno de 1811,* printed in São Paulo by the Imprensa Oficial do Estado in 1936.

______, *Pluto Brasiliensis: Eine Reihe von Abhandlungen über Brasiliens Gold-, Diamanten- und anderen mineralischen Reichthum* . . . (Berlin: G. Reimer, 1833). A Portuguese translation of this work with notes was done by Domício de Figueiredo Murta. Biblioteca Pedagógica Brasileira, ser. 5, Brasiliana, 2 vols. (São Paulo: Companhia Editora Nacional, 1944), vol. 257.

Frézier, Amédée François, *Relation du voyage de la Mer du Sud aux côtes du Chili, du Pérou, et du Brézil, foit pendant les années 1712, 1713 & 1714* (Amsterdam: Chez P. Humbert, 1717). An English edition printed in London in 1717 for J. Bowyer appeared under the title *A Voyage to the South Sea, and Along the Coasts of Chili and Peru, in the Years 1712, 1713, and 1714, Particularly Describing the Genius and Constitution of the Inhabitants, As Well Indians As Spaniards* . . . . A Spanish version entitled *Relación del viaje por el Mar del Sur á las costas de Chile i el Perú durante los años de 1712, 1713 i 1714* was translated by Nicolas Peña M. and published in Santiago, Chile, in 1909 by the Imprenta Mejia.

Gardner, George, *Travels in the Interior of Brazil, Principally through the Northern Provinces, and the Gold and Diamond*

Districts during the Years 1836–41 (London: Reeve, Brothers, 1846).

Gemelli Careri, Giovanni Francesco, *Giro del mondo nuovo* (Venice: Presso G. Malachin a Spese di G. Maffei, 1719). An English edition appeared in A. and J. Churchill's *A Collection of Voyages and Travels,* 3d ed. (London, 1745), with the title *A Voyage around the World.* Two editions have appeared in Spanish. One, translated by José María de Agreda y Sánchez with a prologue by Alberto María Carreño, entitled *Las cosas mas considerables vistas en la Nueva España* (1697) was printed by Ediciones Xochitl in Mexico City in 1946. Another, using the same translation with an introduction by Fernando B. Sandoval, with the title *Viaje a la Nueva España* appeared in two vols. in Mexico City in 1955 printed by Libro-Mex., Editores.

Haenke, Thaddäus, *Viaje por el virreinato del Río de la Plata* (1795), Colección Buen. Aire no. 24 (Buenos Aires: Emecé Editores, S.A., 1943).

———, *Descripción del reyno de Chile,* Agustín Edwards, Introduction (Santiago: Nascimiento, 1942). For a biography of Haenke, see Biographies, the study by Arnade and Kuehnel.

Helms, Anthony Zachariah, *Travels from Buenos Ayres by Potosí to Lima* (London, 1806). This is an English translation of the original German work, *Tagebuch einer Reiser durch Peru* (Dresden, 1798). Helms, a member of Baron Nordenflict's expedition to Potosí, remained behind to offer courses in mining and metallurgy.

Humboldt, Alexander von, *Voyage de Humboldt et Bonpland,* 23 vols. (Paris: F. Schoell, 1805–1834). Part 3 of this study is the famous *Essai politique sur le Royaume de la Nouvelle Espagne,* 2 vols. (Paris: F. Schoell, 1811). A Spanish translation by Vicente González Arnao entitled *Ensayo político sobre el reino de la Nueva-*

España appeared in four volumes printed in Paris by Rosa in 1822. A recent printing with notes and editing by Vito Alessio Robles in five volumes was printed in Mexico by Robredo in 1941. A popular edition edited by Juan A. Ortega y Medina was issued by Editorial Porrua, S.A:, in Mexico City in 1966 as no. 39 of their series "Sepan Cuantos . . . ." An English translation was done by John Black under the title *Political Essay on the Kingdom of New Spain* and printed in London by Longman, Hurst, Rees, Orme & Brown in 1811. John Taylor compiled *Selections* from Humboldt relating to the climate, inhabitants, produce, and mines of Mexico, published in London in 1827. Black's translation has been reprinted in four volumes by AMS Press in New York in 1969. Part 1 of the *Voyage* was translated by Helen Maria Williams in two volumes with the title *Personal Narrative of Travels to the Equinoctial Regions of the New Continent during the Years 1799–1804.* It was published by Longman, Hurst, Rees, Orme & Brown in London; the third edition appeared in 1822. Williams's translation has been reprinted in seven volumes by the AMS Press in New York, 1969.

Juan y Santacilia, Jorge, and Antonio de Ulloa, *Noticias secretas de América, sobre el estado naval, militar, y político de los reynos del Perú y provincias de Quito, costas de Nueva Granada y Chile: Gobierno y régimen particular de los pueblos de Indios: Cruel opresión y estorsiones de sus corregidores y curas: Abusos escandalosos introducidos entre estos habitantes por los misioneros; causas de su origen y motivos de su continuación por el espacio de tres siglos. Escritas fielmente según las instrucciones del excelentisimo señor marqués de estado, y presentadas en informe secreto a s. m. c. et señor don Fernando VI por don Jorge Juan y don Antonio de Ulloa . . . . Sacadas a luz para el verdadero conocimiento del gobierno de los Españoles en la América meridional, por don David Barry,* 2 vols. in 1 (London: Imprenta de R. Taylor, 1826). A recent Spanish edition edited by Gregorio

Weinberg was published in Buenos Aires by Ediciones Mar Océano in 1953 under the title *Noticias secretas de América.*

______, *Relación histórica del viage a la América meridional hecho de órden de S. Mag. para medir algunos grados de meridiano terrestre, y venir por ellos en conocimiento de la verdadera figura, y magnitud de la tierra, con otras varias observaciones astronómicas y phisicas,* 4 vols. (Madrid: A. Marín, 1748). The standard translation into English is by John Adams, *A Voyage to South America: Describing at Large the Spanish Cities, Towns, Provinces, &c. on That Extensive Continent; Undertaken by Command of the King of Spain by Don George Juan, and Antonio de Ulloa,* 2 vols. (London, 1806). A modern abridgement of the Adams translation was done, with a historical introduction, by Irving A. Leonard (New York: A. A. Knopf, 1964).

La Condamine, Charles Marie de, *Journal du voyage fait par ordre du roi, à l'équateur, servant d'introduction historique à la mesure de trois premiers degrés du méridien* (Paris: Imprimerie Royale, 1751). A shorter version had appeared a few years earlier under the title *Relation abrégée d'un voyage fait dans l'intérieur de l'Amérique méridionale depuis la côte de la Mer du Sud jusqu'aux côtes du Brésil & de la Guiane en descendant la rivière des Amazones* (Paris: Veuve Pissot, 1745). A Spanish translation by Frederico Ruíz Morcuende was printed in Madrid by Espasa-Calpe, S.A., in 1935 with the title *Relación abreviada de un viaje hecho por el interior de la América meridional desde la costa del Mar del Sur hasta las costas del Brasil y de la Guayana, siguiendo el curso del río de las Amazonas.* A second edition under the title *Viaje a la América meridional* was printed by Espasa-Calpe Argentina, S.A., in Buenos Aires and Mexico City in 1945.

Las Barras y Aragón, Francisco de, *Un trabajo de don Tadeo Haenke sobre la provincia de Cochabamba.* Ser. B, no.

284. (Madrid: Publicaciones de la Real Sociedad Geográfica, 1952).

Lyon, George F., *A Journal of a Residence and Tour in the Republic of Mexico in the Year 1826, with Some Account of the Mines of That Country,* 2 vols. (London: John Murray, 1828). Lyon surveyed the state of the Mexican mining industry for a group of British investors. There is much interesting detail and many illustrations.

Mawe, John, *Travels in the Interior of Brazil, Particularly in the Gold and Diamond Districts of That Country* (London, 1812). A most astute and informative report.

Novo y Colson, Pedro de, ed., *Viaje político-científico alrededor del mundo por las corbetas descubierta y atrevida al mando de los capitanes de Navío D. Alejandro de Malaspina y D. José de Bustamante y Guerra desde 1789 a 1794* (Madrid, 1885).

Poinsett, Joel R. *Notes on Mexico Made in the Autumn of 1822 Accompanied by an Historical Sketch of the Revolution* (Philadelphia: H. C. Carey and I. Lea, 1824). The author, the first United States Minister to Mexico, had made a brief trip to Mexico three years before his appointment in order to survey the country.

Saint-Hilaire, Augustin François César Provençal de, *Voyage dans le district des diamans et sur le littoral du Brésil, suivi de notes sur quelques plantes charactéristiques . . . ,* 2 vols. (Paris: Librairie Gide, 1833). This section is part of a larger work: *Voyage dans l'intérieur du Brésil,* 8 vols. (Paris, 1830–1851).

Ulloa, Antonio de, *Noticias Americanas: Entretenimientos phísicos-históricos, sobre la América meridional, y la septentrional oriental . . .* (Madrid: Imprenta de F. Manuel de Mena, 1772). A modern edition edited by Luis Aznar was reprinted by Editorial Nova in Buenos Aires in 1944 in its series *Colección Viajeras de las Américas.*

______, *Relación histórica del Viaje a la América Meridional,* 4 vols. (Madrid: A. Marin, 1748). The discovery of platinum in the mines of Chocó was first described in vol. II, ch. 10, no. 1026, p. 606.

Ward, Henry George, *Mexico during the Years 1825, 1826 and Part of 1827: With an Account of the Mining Companies and of the Political Events in That Republic to the Present Day,* 2 vols. 2d ed., enlarged (London: H. Colburn, 1829). A first edition had appeared a year earlier in two volumes under the title *Mexico in 1827.* Ward was first British Minister to Mexico, and this report is a summation of his experiences and travels. It was used to interest British investors in Mexican mines.

Descriptions of Technological Processes

This section is divided into two parts. Part A is a listing of books about mining technology and metallurgy in general. It covers works current during the colonial period to give an idea of the state of the art and industry. It also lists secondary works surveying both the general history of mining and metallurgical practice and the history of science in Spain. Part B consists of a similar listing, but of works directly relating to America.

A. General Descriptions

Agricola, Georg, *De re metallica* (1556), with annotations by Herbert C. and Lou H. Hoover, eds, and trans. (London: The Mining Magazine, 1912); reprinted in 1950 by Dover Publications in New York.

Aitchison, Leslie, *A History of Metals,* 2 vols. (London: Macdonald & Evans, 1960); reprinted in 1960 by Interscience Publishers in New York.

Anales de ciencias naturales, 6 vols. (Madrid, 1799–1804). (The first three issues appeared under the title *Anales de historia natural.*)

Barbier, Marcel N., *Les mines et les arts à travers les âges* (Paris: Société de l'Industrie Minérale, 1956).

Berdegal de la Cuesta, Juan, *Cartilla práctica sobre el laboreo de las minas y reconocimiento y beneficio de los metales* (Madrid, 1839).

Biringuccio, Vannoccio, *De la pirotechnia* (Venice, 1540). An English edition was translated by Cyril S. Smith and Martha T. Grudi as *The Pirotechnia of Vannoccio Biringuccio* (New York: American Institute of Mining and Metallurgical Engineers, 1942).

Blake, William Phipps, *Silver Mines and Silver Ores* (New Haven, Conn., 1861).

Calvo, Felipe A., *La España de los metales: Notas para una historia* (Madrid: Patronato Juan de la Cierva de Investigación Científica y Técnica, Centro Nacional de Investigaciones Metalurgicas, 1964).

Carracido, José A., *Estudios histórico-criticos de la ciencia española* (Madrid, 1897).

Collins, Henry F., *Metallurgy of Lead and Silver,* Griffin's Metallurgical Series (London: C. Griffin & Co., Ltd., 1899–1900).

Elhuyar, Fausto, "Theorie der Amalgamation." *Bergbaukunde,* vol. I (1789), pp. 238–263; vol. II (1790), pp. 200–298.

Launay, Louis de, *La conquête minérale,* Bibliothèque de Philosophie Scientifique (Paris: E. Flammerion, 1908).

Menéndez y Pelayo, Marcelino, *La ciencia Española,* vol. III of *Obras Completas* 4th ed., rev. (Madrid, 1918).

Mieli, Aldo, *Panorama general de historia de la ciencia, Serie menor: Historia y filosofía de la ciencia,* 8 vols. in 5, 2d ed. (Buenos Aires: Espasa-Calpe, S.A., 1950–1961).

Moles, Enrique, *El momento científico español (1775–1825)* (Madrid, 1934).

Parsons, William Barclay, *Engineers and Engineering in the Renaissance* (Baltimore: The Williams and Wilkins Co., 1939).

Partington, James R., *Origins and Development of Applied Chemistry* (London and New York: Longmans, Green and Co., 1935).

______, *A Short History of Chemistry* (New York: The Macmillan Co., 1937).

Rey Pastor, Julio, *La ciencia y la técnica en el descubrimiento de América,* Colección Austral 301 (Buenos Aires and Mexico: Espasa-Calpe Argentina, S.A., 1942).

Rickard, Thomas A., *Man and Metals: A History of Mining in Relation to the Development of Civilization,* 2 vols. (New York: Whittlesey House, McGraw-Hill Book Company, 1932).

Weeks, Mary Elvira, *Discovery of the Elements,* 6th ed. (Easton, Pa.: Journal of Chemical Education, 1956).

B. New World Techniques

Barba, Alvaro Alonso, *Arte de los metales en que se enseña el verdadero beneficio de los de oro, y plata por asogue; el modo de fundirlos todos, y como se han de refinar y apartar unos de otros* (Madrid: Imprenta del Reyno, 1640). This edition was ordered destroyed by the Inquisition, and only three copies, now in the British Museum, have survived. There followed an English translation by Lord Edward Montagu (London: Mearne, 1670); another English edition in 1674 (London: Mearne); a second Spanish edition in 1675 (Córdova); an Italian edition in 1675; the first German edition in 1676 (Hamburg: Schultens); and still another Spanish one in 1680. In the eighteenth century there were Spanish editions in 1729, 1768, and 1770; English editions in 1738, 1739, and 1749; German ones in 1726, 1749, and 1767; French ones in 1729, 1730, 1751, another in 1750–1751, and 1752; and Dutch ones in 1735, 1740, and

1752. In the nineteenth century the work was printed in Madrid in 1811, 1842, and 1852; in Lima in 1817 and 1842–1843; and in Santiago (Chile) in 1878. In the present century we have an English translation by Ross E. Douglass and E. P. Mathewson under the title *El Arte de los metales (Metallurgy),* printed by John Wiley & Sons, Inc., New York, in 1923; and the facsimile edition issued in 1925, in Mexico City, by the Compañía Fundidora de Fierro y Acero de Monterrey, S.A. from the Madrid edition of 1770, printed by the Viuda de Manuel Fernández. The plates that served for this edition were presented to the School of Mining Engineering in Madrid, which used them to print a facsimile edition in 1932. The latest edition is the recent one (1967) published in the city of Potosí, Bolivia, in the very place where Padre Alvaro Alonso Barba wrote it 330 years before. This magnificent edition was prepared, edited, and prefaced by Don Armando Alba (Bibliographic Note by C. Prieto).

Bargalló, Modesto, "Método de beneficio de los metales de oro y plata por fundición del mexicano José Garcés y Eguía," *Ciencia* (Mexico), vol. XII (1952), pp. 155–159.

———, "Primer tratado completo y sistemático del beneficio de patio de don José Garcés y Equía, metalúrgico mexicano," *Ciencia,* vol. XII (1952), pp. 199–206.

———, "La amalgamación de menas de plata en Nueva España en la segunda mitad del siglo XVI," *Ciencia,* vol. XV (1955), pp. 216–218.

———, "Las investigaciones de Fausto de Elhuyar sobre la amalgamación de menas de plata," *Ciencia,* vol. XV (1955), pp. 261–264.

———, "El beneficio de amalgamación de patio;

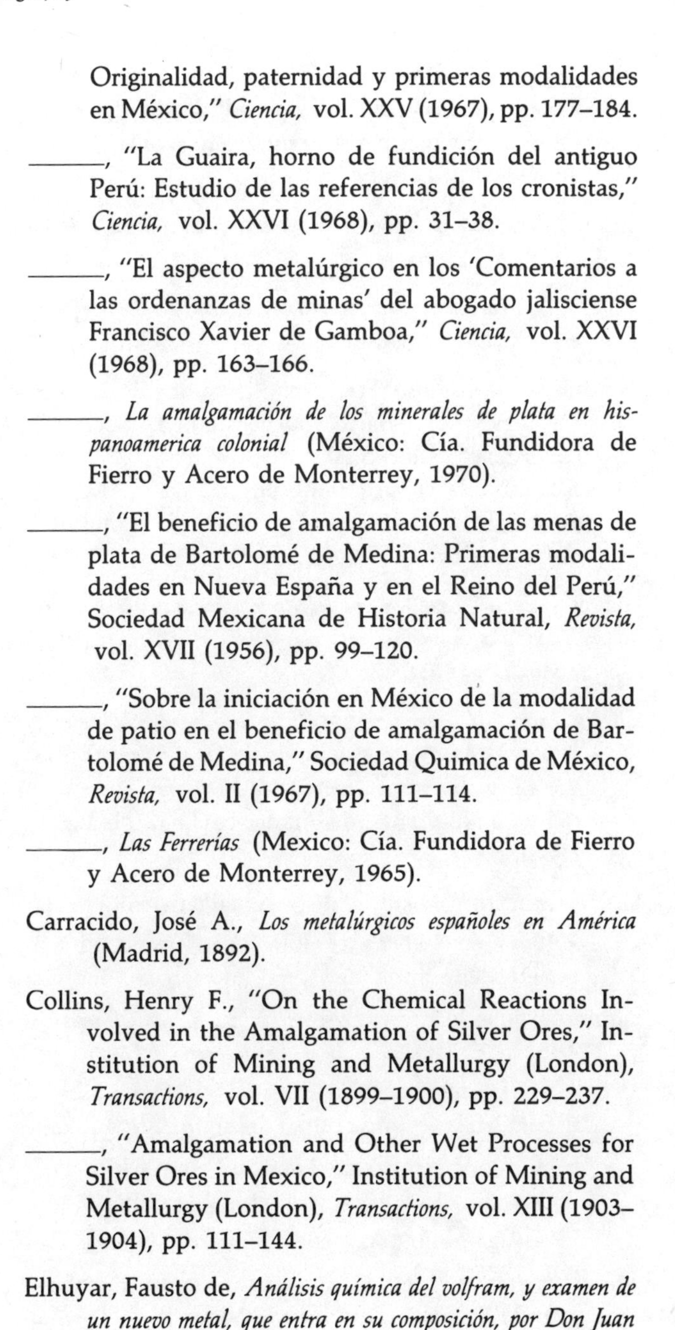

Originalidad, paternidad y primeras modalidades en México," *Ciencia,* vol. XXV (1967), pp. 177–184.

———, "La Guaira, horno de fundición del antiguo Perú: Estudio de las referencias de los cronistas," *Ciencia,* vol. XXVI (1968), pp. 31–38.

———, "El aspecto metalúrgico en los 'Comentarios a las ordenanzas de minas' del abogado jalisciense Francisco Xavier de Gamboa," *Ciencia,* vol. XXVI (1968), pp. 163–166.

———, *La amalgamación de los minerales de plata en hispanoamerica colonial* (México: Cía. Fundidora de Fierro y Acero de Monterrey, 1970).

———, "El beneficio de amalgamación de las menas de plata de Bartolomé de Medina: Primeras modalidades en Nueva España y en el Reino del Perú," Sociedad Mexicana de Historia Natural, *Revista,* vol. XVII (1956), pp. 99–120.

———, "Sobre la iniciación en México de la modalidad de patio en el beneficio de amalgamación de Bartolomé de Medina," Sociedad Quimica de México, *Revista,* vol. II (1967), pp. 111–114.

———, *Las Ferrerías* (Mexico: Cía. Fundidora de Fierro y Acero de Monterrey, 1965).

Carracido, José A., *Los metalúrgicos españoles en América* (Madrid, 1892).

Collins, Henry F., "On the Chemical Reactions Involved in the Amalgamation of Silver Ores," Institution of Mining and Metallurgy (London), *Transactions,* vol. VII (1899–1900), pp. 229–237.

———, "Amalgamation and Other Wet Processes for Silver Ores in Mexico," Institution of Mining and Metallurgy (London), *Transactions,* vol. XIII (1903–1904), pp. 111–144.

Elhuyar, Fausto de, *Análisis química del volfram, y examen de un nuevo metal, que entra en su composición, por Don Juan*

Josef y Don Fausto de Elhuyar (Vitoria: G. H. Robles y Revilla, 1783). An English edition of this first essay on the discovery of tungsten (wolfram) was published in London in 1785.

———, *Indagaciones sobre la amonedación` en Nueva España* (Madrid, 1818).

———, *Informe sobre la diferencia entre el beneficio por azogue y de la fundición* (1793).

Fernández del Castillo, Francisco, "Algunos documentos nuevos sobre Bartolomé de Medina," Sociedad Científica "Antonio Alzate" (Mexico), *Memorias y Revista,* vol. XLVII (1927), pp. 207–251. Also see the article on Medina under Alan Probert in Biographies below.

Garcés y Eguía, José, *Nueva teórica y práctica del beneficio de los metales de oro y plata por fundición y amalgamación* (Mexico, 1803).

García, Trinidad, *Memoria leida en la Sociedad de Geografia y Estadística Mexicana en la sesion del día 13 de octubre de 1888 sobre la teoría del beneficio de amalgamación por patio* (Mexico: Imprenta de Francisco Díaz de León, 1888).

Greve, Ernesto, "Historia de la amalgamación de la plata," *Revista Chilena de Historia y Geografía,* vol. CII (1943), pp. 158–259.

Izquierdo, José Joaquin, *La primera casa de las Ciencias en México: el Real Seminario de Minería, 1792–1811* (Mexico: Ediciones Ciencia, 1958).

Lohmann Vîllena, Guillermo, "Enrique Garcés, descubridor del mercurio en el Peru, poeta y arbitrista," *Anuario de estudios americanos,* vol. V (Seville, 1948), pp. 439–482.

Meseguer Pardo, Jose, "El esfuerzo minero y metalúrgico de España en el Nuevo Mundo," *Boletin de la*

Real Sociedad Geográfica, vol. LXXXV (July–September 1949), pp. 339–369.

Monardes, Dr. Nicolás, *Dos libros: El uno trata de Todas las cosas q̃ traê de ñras Indias Occidẽtales, que sirven al vso de Medicina* (Seville: Impresso en Casa de Sebastian Trujillo, 1565). A modern version with the title *Dialogo del hierro* was published in Mexico City, 1961, by the Compañía Fundidora de Fierro y Acero de Monterrey. The first translation into English was done by Frampton in 1577 from the Spanish edition of 1574 under the title *Ioyfull Newes ovt of the Newe-founde Worlde. . . .* It was published in London in 1577 for W. Norton. *The Dialogue of Yron, Which Treateth of the Greatness Thereof, and How It Is the Most Excellent Metall of All Others* was included by John Frampton in his *Ioyfull Newes out of the New-founde Worlde* (London, 1596). A modern edition, with an introduction by Stephen Gaselee, was published under the title *Joyfull Newes out of the Newe Founde Worlde* in two volumes by A. A. Knopf, 1925, in New York.

Motten, Clement G., *Mexican Silver and the Enlightenment* (Philadelphia: University of Pennsylvania Press, 1950).

Mutis, Juan Celestino, *Relación de las operaciones y experimentos para indagar cuál es el mejor método de beneficiar los minerales, si el de la fundición o el de la amalgamación* (1786).

Oñate, Juan de, *Tratado . . . de re metálica,* in Zerón Zapata, Miguel., ed. *La puebla de Los Angeles en el siglo XVII* (Mexico: Ed. Patria, 1945).

Orozco, Juan Manuel de, *Cartilla metálica, que enseña desde sus principios a conocer y beneficiar toda suerte de metales, y nueva quema de ellos, y algunos materiales de que se pueden usar para limpiar la plata con rara felicidad y certidumbre y mucho aumento, y menos pérdida de azogue* (Lima, 1737).

Paoli, Umberto G., *L'etá aurea della metallurgia ispano-coloniale. Quaderni di storia della scienza,* no. 10 (Rome: Casa Editrice L. da Vincio, 1927). This work also appeared in the *Archivo di Storia della Scienza,* vol. VII (1926), pp. 95–115 and 226–234; vol. VIII (1927), pp. 83–94, 210–213, 364–376, 496–498.

———, "Il Metallurgista Spagnolo Alvaro Alonso Barba da Villa Lepe (1569–1662)," *Archivio di Storia della scienza,* vol. III (1922), pp. 150–168.

Rio, Andrés del, *Los elementos de orictognosia ó del conocimiento de los fósiles . . . para el uso del Real Seminario de Minería de Mexico* (Mexico: Mariano Joseph de Zúñiga, 1795).

Rodríguez Carracido, José, *Los metalúrgicos españoles en América,* Ateneo de Madrid, Conferencias Públicas, no. 16 (Madrid: Establecimiento Tipográfica "Sucesores de Rivadeneyra," 1892).

Sarria, Francisco Xavier de, *Ensayo de metalurgia, o descripción por mayor de las catorce materias metálicas, del modo de ensayarlos, del laborio de las minas, y del beneficio de los frutos minerales de la plata* (Mexico: Impreso por D. Felipe de Zuñiga y Ontiveros, 1784).

———, *Suplemento al ensayo de metalurgia* (Mexico: Por Don Felipe de Zuñiga y Ontiveros, 1791).

Sonneschmidt, Friedrich T. (Federico Sonneschmid), *Tratado de amalgamación de Nueva España* (1805), sacado a luz por d. J.M.F. (Mexico: Librería de Bossange (padre), Antoran y Cía, 1825).

Tejado Fernández, Manuel, "Un informe de Ulloa Sobre la explotación del platino," *Saitabi,* vol. VII, nos. 31–32 (1949), pp. 51–76.

Torre Barrio y Lima, Lorenzo Felipe de la, *Arte y Cartilla del nuevo beneficio de la plata en todo genero de metales frios y calientes* (Lima: Antonio Joseph Gutiérrez de Zevellos, 1738).

Vásquez Machicado, Humberto, "En torno a la alquimia del Padre Barba," *Universidad de San Francisco Xavier* (Sucre), vol. XVI, nos. 39 and 40 (1951), pp. 362–381.

Whitaker, Arthur P., "The Elhuyar Mining Missions and the Enlightenment," *Hispanic American Historical Review,* vol. XXXI (1951), pp. 558–585.

Zavala, Silvio, "La amalgamación en la minería de la Nueva España," *Historia Mexicana,* vol. XI, no. 3 (1962), pp. 416–421.

Mines and Mining and Their Economic History: Secondary Sources

This section lists secondary works of a diverse nature, but all entries contain material pertaining to mining in colonial Hispanic America. Because there is not yet an equivalent in English, the reader is referred to Modesto Bargalló's *La minería y metalurgia en la América Española* as the best general introduction to the topic. For an introduction in English, see Bailey Diffie, *Latin American Civilization: Colonial Period* (Harrisburg, Pa.: Stackpole Sons, 1945); C. Haring, *Spanish Empire in America* (New York: Oxford Press, 1947), reprinted by Harcourt, Brace and World, New York 1963; Charles Boxer, *The Golden Age of Brazil* (Berkeley: The University of California Press, 1962); and, as listed in this section, the studies by Cleland, Fox, and Jenison.

Aguilera, José G., et al., *El mineral de Pachuca,* Instituto Geológico de Mexico, *Boletín*, nos. 7–9 (Mexico: Oficina Tipográfica de la Secretaría de Fomento, 1897).

Alvarez, Manuel F., *El Palacio de Minería* (Mexico, 1910).

André, Marius, "Le Baron de Nordenflicht . . . et les mineurs allemands au Pérou," *Revue de l'Amérique Latine*, vol. VIII (1924), 289–306.

Arana, Pedro P., *Las minas de azogue del Perú* (Lima: Imprenta de "El Lucero," 1901).

Artola y Guardiola, Antonio, "Notas para una historia de la Imperial Villa de Potosí," unpublished doctoral dissertation, University of Madrid (1909).

Bakewell, P. J., *Silver Mining and Society in Colonial Mexico, 1546–1700,* Cambridge Latin American Series, no. 15 (Cambridge: The University Press, 1971).

Ballesteros Gaibrois, Manuel, *Descubrimiento y fundación del Potosí* (Zaragoza, Spain: Seminario Nacional de Hispanidad, Delegación de Distrito de Educación Nacional, 1950).

Bargalló, Modesto, "Alvaro Alonso Barba: Su 'Arte de los Metales,' " *Ciencia* (Mexico), vol. XXVII, no. 2 (1969).

______, "Bartolomé de Medina y el beneficio de patio: Algunos aspectos poco conocidos." Sociedad Qúimica del Perú, *Boletín* (Lima), vol. XVIII (1952), pp. 101–108.

______, "Minas prehispanicas y coloniales," *Diccionario Porrúa de historia, biografía y geografía de México, Supplemento* (Mexico: Editorial Porrúa, 1966).

______, *La minería y la metalurgia en la América Española durante la época Colonial* (México: Fondo de Cultura Económica, 1955).

______, *La química inorganica y el beneficio de los metales en el México prehispanico y colonial* (Mexico: Facultad de Quimica, Universidad Nacional Autónoma de México, 1966).

Bastani, Tanaus Jorge, *Minas e Minérios no Brasil: Tesouros, cidades préhistóricas e minas abandonadas* (Rio de Janeiro: Livraria Freitas Bastos, 1957).

Bernstein, Marvin D., *Mexican Mining Industry, 1890–1950: A Study of the Interaction of Politics, Economics and Technology* (Albany, N.Y.: State University of New York Press, 1964).

Berry, Edward W., and Singewald, Joseph T., Jr., *The Geology and Paleontology of the Huancavelica Mercury District* (Baltimore: The Johns Hopkins Press, 1922).

Bertran de Quintana, Miguel, "El Real Seminario de Minería y Velásquez de León, Elhuyar, y del Rio," *Excelsior* (Mexico, Feb. 22, 1935).

Brading, David A., "Mexican Silver Mining in the Eighteenth Century: The Revival of Zacatecas," *Hispanic American Historical Review*, vol. L, no. 4 (November 1970), pp. 665–681.

———, *Miners and Merchants in Bourbon Mexico, 1763–1810,* Cambridge Latin American Series, no. 10 (Cambridge: The University Press, 1971). One of the most important recent works, both for originality of content and thorough bibliography.

Calógeras, João Pandiá, *Formação histórica do Brasil,* 3d. ed. (São Paulo, 1938).

Cardozo, Manoel, "The Brazilian Gold Rush," *The Americas* (Franciscan Academy), vol. III, no. 2 (October 1946), pp. 137–160.

———, "Alguns subsídios para a história da cobrança do quinto na capitania de Minas Gerais até 1735" (Lisbon, 1938). Reprinted from *I Congreso da História da expansão Portugesa no Mundo*, III Secção (Lisbon, 1937).

———, "Dom Rodrigo de Castel–Blanco and the Brazilian El Dorado, 1673–1682," *The Americas* (Franciscan Academy), vol. I, no. 2 (October 1944), pp. 131–159.

Carranza Sánchez, Fortunado, "Sinopsis histórico-científica de la industria minera en el Perú a través de los períodos incaico y colonial," *Revista Universitaria* (Lima), vol. XVII (1923), part II, pp. 166–230. This study was also printed separately (Lima: Torres Aguirre, 1922).

Carvalho Franco, Francisco de Assis, *História das minas de São Paulo* (São Paulo, 1964).

Castillo, Antonio del, *Memoria sobre las minas de azogue de América* (Mexico, 1871). This is a revision of a report written in 1845 dealing mainly with Mexico and California.

Cleland, Robert G., "The Mining Industry of Mexico: A Historical Sketch," *Mining and Scientific Press*, vol. CXXIII (July 2, 1921), pp. 13–20; (Nov. 5, 1921), pp. 638–642.

Cobb, Gwendolin B., "Potosí, a South American Mining Frontier," in *Greater America: Essays in Honor of Herbert Eugene Bolton* (Berkeley: The University of California Press, 1943), pp. 39–58.

______, "Potosí and Huancavelica: Economic Bases of Peru, 1545 to 1640," unpublished doctoral dissertation, University of California, Berkeley (1947).

______, "Supply and Transportation for the Potosí Mines, 1545–1640," *Hispanic American Historical Review*, vol. XXIX (1949), pp. 25–45.

Conga Argüelles, José, *Diccionario de hacienda, con aplicación a España . . .*, 2 vols. 2d ed. (Madrid, 1833–1834). Notes on mines, especially Real del Monte.

Crespo y Martinez, Gilberto, *México, industria minera: Estudio de su evolución* (Mexico: Oficina Tipográfica de la Secretaría de Fomento, 1903).

Dahlgren, Charles B., *Historic Mines of Mexico: A Review of Mines of That Republic for the Last Three Centuries* (New York: Printed for the author, 1883). A translation, authorized by the Sociedad Mexicana de Minería, was published under the title *Minas históricas de la República Mexicana* (Mexico: Oficina Tipográfica de la Secretaría de Fomento, 1887). There is also a guide entitled *Handbook to the "Historic Mines of Mexico"* (New York, 1886).

Derby, Orville A., "Os primeiros descobrimentos de ouro em Minas Gerais," *Revista do Instituto Histórico e Geográfico de São Paulo,* vol. 5 (1899–1900), pp. 240–278.

———, "Os primeiros descobrimentos de ouro nos distritos de Sabará e Caeté," *Revista do Instituto Histórico e geográfico de São Paulo,* vol. 5 (1899–1900), pp. 279–295.

Deustua Pimentel, Carlos, "La Expedición mineralogista del Barón Nordenflicht al Perú, " *Mercurio Peruano* (Lima), vol. XXXVIII (October–November 1957), 510–519.

Diffie, Bailey W., "Estimates of Potosí Mineral Production, 1545–1555," *Hispanic American Historical Review,* vol. XX no. 2 (May 1940), pp. 275–282.

Duport, St. Clair, *De la production des métaux précieux au Mexique* (Paris: F. Didot frères, 1843).

Fernández, Justino, *El palacio de minería de México*, Ediciones del IV Centenario de la Universidad de México, Serie Conmemorativa, no. 4. (Mexico: Instituto de Investigaciones Estéticas, 1951).

Fox, John, "The Beginnings of Spanish Mining in America: The West Indies and Castilla del Oro," unpublished doctoral dissertation, University of California, Berkeley (1940).

García, Trinidad, *Los mineros Mexicanos: Colección de artículos sobre tradiciones y narraciones mineras, descubrimiento de las minas más notables, fundición de las poblaciones minerales más importantes y particularmente sobre la crisis producida por la baja de la plata* (Mexico: Oficina Tipográfica de la Secretaría de Fomento, 1895).

Garfias, V. R., "Historical Outline of Mineral Production in Mexico," American Institute of Mining and Metallurgical Engineers, *Transactions*, CXXVI (1937), pp. 346–355.

Gicklhorn, Renée, *Die Bergexpedition des Freiherrn von Nordenflycht und die deutschen Bergleute in Peru*, Freiberger Forschungshefte, no. 40 (Leipzig: Reihe D., Kultur und Technik, 1963).

______, and Gicklhorn, Josef, *Deutsche Arbeit bei der bergmännischen Erschliessung der Vicekönigreiche Peru und La Plata von Alexander v. Humboldt*, Geografische Gesellschaft, *Mitteilungen der* . . . , vol. XLVIII (Hamburg, 1944).

______, *Th. Haenkes Bedeutung für die Erforschung Südamerikas von Alexander v. Humboldt*, Geographische Gesellshaft, *Mitteilungen der* . . . , vol. XLVII (Hamburg, 1941).

González Reyna, Jenaro, *Memoria geológico-minera del estado de Chihuahua (minerales metálicos),* Contribución al XX Congreso Geológico Internacional del Instituto de Geología de la Universidad Autónoma de México (Mexico: Editorial Stylo, 1956).

______, "Los períodos de desarrollo de la minería mexicana," *Memoria del Congreso Cientifico Mexicano: Ciencias físicas y matemáticas* (Mexico: Universidad Nacional Autónoma de México, 1953), vol. III, pp. 277–303.

Grote, D. Alberto, *Adelantados de la minería en México durante el siglo del centenario de Independencia*, Series: Concurso Científico y Artistico del Centenario (Mexico, 1911). This article also appeared in the Instituto Mexicano de Minas y Metalurgía, *Informes y memorias*, vol. II (1912), pp. 51–61, 67–80.

Hamilton, Earl J, "Imports of American Gold and Silver into Spain, 1503–1660," *Quarterly Journal of Economics,* vol. XLIII (1929), pp. 436–472.

Hanke, Lewis, *La "Historia de la Villa Imperial de Potosí" de Bartolomé Arzáns de Orsúa y Vela,* Alberto Tauro, Preface (Lima: Editorial de la Universidad Nacional Mayor de San Marcos, n.d.).

______, *The Portuguese in Spanish America: With Special Reference to the Villa Imperial de Potosí;* reprinted from *Revista de Historia de América,* no. 51 (1961) n. p.

______, *The Portuguese and the Villa Imperial de Potosí,* offprint from *Actas do III Colóquio Internacional de Estudos Luso-Brasileiros,* vol. II. (Lisbon, 1960).

______, *La Villa Imperial de Potosí: Un capítulo inédito en la historia del Nuevo Mundo,* Gunnar Mendoza, trans. (Sucre: Universidad de San Francisco Xavier, 1954).

______, *The Imperial City of Potosí: An Unwritten Chapter in the History of Spanish America* (The Hague: Martinus Nijhoff, 1956).

Haring, Clarence H., "American Gold and Silver Production in the First Half of the Sixteenth Century," *Quarterly Journal of Economics,* vol. XXIX (May 1915), pp. 433–479.

Hawley, C. E., "Notes on the Quicksilver Mine of Santa Barbara in Peru: Condensed from a Report Made to the New Almaden Quicksilver Co," *American Journal of Science and Arts,* 2d ser., vol. XLV (1868), pp. 5–9.

Helmer, Marie, "Edelmetalle Perus in der Kolonialzeit," *Saeculum,* vol. XIII (1962), pp. 293–300.

Hermann, Alberto, *La produccion de oro, plata i cobre en Chile desde los primeros dias de la conquista hasta fines de agosto de 1894* (Santiago de Chile, 1894).

Howe, Walter, *The Mining Guild of New Spain and Its Tribunal General, 1770–1821* (Cambridge, Mass.: Harvard University Press, 1949).

"Informação sobre as minas do Brasil," *Anais da Biblioteca Nacional de Rio de Janeiro,* vol. LVII (1935), pp. 155–186.

Instituto Mexicano de Minas y Metalurgía, *Informes y Memorias* (Mexico: 1910–1912).

Jenison, H. A. C., "The Mining History of Mexico," *Engineering and Mining Journal,* vol. CXV (Feb. 24, 1923).

Jiménez, Carlos P. "Reseña histórica de la minería en el Perú," Ministerio de Fomento (Perú), *Síntesis de la minería peruana en el centenario de Ayacucho* (Lima, 1924), vol. I, pp. 1–71.

Launay, C., "Mines et Industries Minières," in R. Bonaparte, L. Bourgeois, et al., eds., *Le Mexique au Débût du XXe Siecle,* 2 vols. (Paris: C. Delagrave, 1904).

Lee, Theophilus Henry, "A Historical Sketch of the Development of Mining in Brazil," *Museu Nacional do Rio de Janeiro, Archivos,* vol. XXII (1919), pp. 197–220.

Lima, Augusto de, Jr., *As primeiras vilas do ouro* (Belo Horizonte, 1962).

———, *História dos diamantes nas Minas Gerais* (Lisbon and Rio de Janeiro, 1945).

Lohmann Villena, Guillermo, *Las minas de Huancavelica en los siglos XVI y XVII* (Seville, 1949).

López Monroy, P, "Apuntes históricos relativos al descubrimiento de las minas de Guanajuato," *El Minero Mexicano,* vol. XXXIV, nos. 7–8 (1899).

Mariscal Romero, Pilar, *Los bancos de rescate de platas,* Publicaciones de la Escuela de Estudios Hispano-Américanos de Sevilla, no. 153 (Seville: Banco de España, 1964).

Marmolejo, Lucio, *Efemérides guanajuatenses: ó, datos para formar la historia de la ciudad de Guanajuato.* 4 vols. (Guanajuato: F. Rodríguez, 1883–1884).

Mata Machado Filho, Aires da, *Arraial do Tijuco, cidade Diamantina* (Rio de Janeiro, 1945).

Mecham, John Lloyd, "The Real de Minas as a Political

Institution," *Hispanic American Historical Review,* vol. VII (February 1927), pp. 45–83.

Mendizabal, Miguel O. de, "Los minerales de Pachuca y Real de Monte en la época colonial," *El Trimestre económico,* vol. VIII (1940), pp. 253–309.

Mill, Nicholas, *A History of Mexico from the Spanish Conquest to the Present Era . . . Also Observations on the Working of Mines,* (London: Printed for Sherwood, Jones and Co., 1824).

Nesmith, R. S., *The Coinage of the First Mint of the Americas at Mexico City, 1536–1572* (New York: American Numismatic Society, 1955).

Noriega, A., "Las minas metálicas del Perú y sus métodos de explotación," *Síntesis de la minería peruana en el centenario de Ayacucho.* (Lima, 1924), vol. I.

Omiste, Modesto, *Crónica Potosina,* 2 vols. (La Paz, Bolivia, 1919).

______, *Memorias históricas sobre acontecimientos políticos occuridos en Potosí en 1810* (Potosí, 1871).

______, *Memorias historicas sobre los acontecimientos occuridos en Potosí en 1811* (Potosí, 1878).

Ordóñez, Ezequiel and Rangel, Manuel, *El Real del Monte.* Instituto Geológico de México, *Boletín,* no. 12 (Mexico: Oficina Tipográfica de la Secretaría de Fomento, 1899).

Oviedo, Basilio Vicente de, *Cualidades y riquezas del Nuevo Reino de Granada* (Bogotá, 1930).

The Pachuca Mining District (Mexico: Compañía de Real del Monte y Pachuca, 1936).

Palacio Atard, Vicente, "El asiento de la mina de Huancavelica en 1779," *Revista de Indias* (Madrid), vol. V (1944), pp. 611–630.

Pederson, Leland R., *The Mining Industry of the Norte Chico, Chile,* National Research Council, Division of

Earth Sciences, Foreign Field Research Program, Report no. 29 (Evanston, Ill.: Department of Geography, Northwestern University, 1966).

Perez, Aquiles R., *Las minas en la Real Audiencia de Quito* (Quito, 1947).

Polo, José Toribio, *Reseña historica de la minería en el Perú* (Lima: Imprenta de Artes Gráficas, 1911).

Powell, Philip W., "The Forty-Niners of Sixteenth Century Mexico," *Pacific Historical Review,* vol. XIX (1950), pp. 235–249.

———, "Franciscans on the Silver Frontier of Old Mexico," *Americas* (Franciscan Academy), vol. III, no. 3 (January 1947), pp. 295–310.

———, *Soldiers, Indians, and Silver: The Northward Advance of New Spain, 1550–1560* (Berkeley: University of California Press, 1952).

Pradeau, Alberto Francisco, *Don Antonio de Mendoza y la Casa de Moneda de México en 1543: Documentos inéditos,* Biblioteca Histórica Mexicana de Obras Inéditas, no. 23 (Mexico: Antigua Librería Robredo de J. Porrúa, 1953).

———, *Numismatic History of Mexico from the Pre-Columbian Epoch to 1823* (Los Angeles: Western Printing Co., 1938).

Quesada, Vicente G., *Crónicas potosinas: Costumbres de la edad medieval hispano-americana,* 2 vols. (Paris, 1890).

Raimondi, A(ntonio), *Minerales del Perú* (Lima, 1878).

Ramirez, Santiago, *Datos para la historia del Colegio de Minería* (Mexico, 1890).

———, *Datos para la historia del Colegio de Minería recogidos y compilados bajo la forma de efemérides* (Mexico: Imprenta del Gobierno Federal en el ex-Arzobispado, 1894). This colegio was called the "Real Seminario de Minería" in the Ordinances of 1783.

———, *Noticia histórica de la riqueza minera de México* (Mexico: Oficina Tipográfica de la Secretaría de Fomento, 1884).

Restrepo, Vicente, *Estudio sobre las minas de oro y plata de Colombia* 2d ed. (Bogotá, 1888).

Ribera, José Bernardez de, Conde de Santiago de la Laguna, *Descripción breve de la muy noble, y leal ciudad de Zacatecas* (Testimonio de Zacatecas) (Mexico: Impressa por J. B. de Hogal, 1732). There is a reprint of this study: Edición de la "Crónica municipal" (Zacatecas, Mexico: Imprenta de la Penitenciaría, 1883).

Rivera, Manuel, *Memoria sobre el mineral de Pachuca* (Mexico, 1864).

Rivero y Ustariz, Mariano Eduardo de, "Memoria sobre la mina de Azogue de Huancavelica y la de Chonta," *Colección de memorias científicas, agrícolas e industriales,* 2 vols. (Brussels, 1857), vol. II, pp. 85–176.

Rivet, Paul, and Arsandaux, H., *La métallurgie en Amérique précolombienne* (Paris: Université de Paris, Travaux et Mémoires de l'Institut d'Ethnologie), vol. XXXIX, 1946.

Rojas, Casto, "El cerro rico de Potosí," *II Congreso Internacional de Historia de América,* 6 vols. (Buenos Aires, 1938), vol. III, pp. 145–158.

Rudolph, William E., "The Lakes of Potosí," *The Geographical Review,* vol. XXVI (1936), pp. 529–554.

Rusconi, Carlos, "Activades mineras antiguas de Mendoza (Argentina)," *La Prensa* (Buenos Aires), Oct. 15, 1967.

Santos, Felicio dos, *Memórias do Distrito Diamantino da Comarca de Sêrro Frio,* 3d. ed. (Rio de Janeiro, 1958).

Serrano, Gustavo P., *La minería y su influencia en el progreso y desarrollo de México* (Mexico, 1951).

Southworth, John R., comp., *Las minas de México* (Liverpool: Blake & Mackenzie, 1905).

Tamayo, Jorge L., "La minería de Nueva España en 1794," *Trimestre económico,* vol. X (1941), pp. 287–319.

Teixeira Coelho, José João, "Observations on Life in Minas Gerais during the Gold Mining Era," in E. Bradford Burns, ed., *A Documentary History of Brazil* (New York: A. A. Knopf, 1967). This excerpt was translated from the *Revista do Instituto Histórico e Geográfico do Brasil,* vol. XV, no. 3 (1852).

Toussaint, Manuel, *Tasco: Su historia, sus monumentos, caracteristicas actuales y posibilidades turisticas* [Mexico: Editorial "Cultura," 1931 (1932)].

Vandelli, Domingos, "Sobre as minas de ouro do Brasil," *Anais da Biblioteca Nacional de Rio de Janeiro,* vol. XX (1898), pp. 266–278.

———, "Sobre os diamantes do Brasil," *Anais da Biblioteca Nacional de Rio de Janeiro,* vol. XX (1898), pp. 279–282.

Velarde, Carlos E., *Notas sobre la minería en el Perú* (1908).

Viana Passos, Zoroastro, *Em tôrno da história de Sabará,* 2 vols. (Belo Horizonte, 1940–1942).

Vignale, Pedro J., *La Casa real de moneda de Potosí* (Buenos Aires: Ediciones de arte "Albatros," 1944).

Wagner, Henry R., "Early Silver Mining in New Spain," *Revista de Historia de América,* no. 14 (June 1942), pp. 49–71.

West, Robert C., *Colonial Placer Mining in Colombia,* Social Science Series no. 2. (Baton Rouge: Louisiana State University Press, 1952).

———, *The Mining Community in Northern New Spain: The Parral Mining District,* Ibero-Americana no. 30 (Berkeley: University of California Press, 1949).

Whitaker, Arthur Preston, *The Huancavelica Mercury Mine* (Cambridge, Mass.: Harvard University Press, 1941).

Wiley, John M., *Historical Report on the Principal Mines in the State of Guanajuato* (Philadelphia, 1903).

Wright, Irene A., "Origenes de la minería en Cuba, 1530–1647," *Reforma Social* (Havana), vol. VII, pp. 450–62; vol. XV, pp. 209–221.

Modern Economic History

The studies in this section are secondary works stressing economic topics. Besides taking up the role of mining in the economy, they deal with the processing, distribution, and influence of precious metal bullion from labor problems to the operations of the mint and Royal Exchequer to the organization of international trade.

Aiton, Arthur S., "Real Hacienda in New Spain under the First Viceroy," *Hispanic American Historical Review,* vol. VI (1926), pp. 232–245.

______, and Wheeler, Benjamin W., "The First American Mint," *Hispanic American Historical Review,* vol. XI (1931), pp. 198–215.

Artiñano y de Galdácono, Gervasio, *Historia del comercio con las Indias durante el dominio de los Austrias* (Barcelona: Talleres de Oliva de Vilanova, 1917).

Azevedo, J. Lucio, *Épocas de Portugal económico* (Lisbon: Teixeira, 1929).

Bancora Cañero, Carmen, "Las remesas de metales preciosos desde El Callao a España en la primera mitad del siglo XVIII," *Revista de Indias,* vol. XIX (1959), pp. 35–85.

Barrio Lorenzot, Juan Francisco de, *El trabajo en México durante la época colonial: Ordenanzas de los Gremios de la Nueva España* (Mexico: Secretaría de Industria, Comercio y Trabajo, 1920).

Borah, Woodrow, *Early Colonial Trade and Navigation between Mexico and Peru* (Berkeley: University of California Press, 1954).

Brito, Lemos, *Pontos de partida para a historia econômica do Brasil,* 2d ed. (São Paulo, 1939).

Chaunu, Pierre, *L'Amérique et les Amériques,* Collection Destins du Monde, no. 8 (Paris: A. Colin, 1964).

Chaunu, Huguette, and Chaunu, Pierre, *Séville et l'Atlantique, 1504 à 1650,* 8 vols. (Paris: A. Colin; S.E.V.P.E.N., 1955–1960).

Crespo Rodas, A., "La 'mita' de Potosí," *Revista Histórica* (Lima), vol. XXII (1955–1956), pp. 169–182.

Del Mar, Alexander, *History of Monetary Systems: A Record of Actual Experiments in Money Made by Various States of the Ancient and Modern World, as Drawn from Their Statutes, Customs, Treatises, Mining Regulations, Jurisprudence, History, Archaeology, Nummulary Systems, Coins, and Other Sources of Information* (London and Chicago: C. H. Kerr & Company, 1896). This volume was reprinted by Augustus M. Kelly in New York in 1969.

———, *A History of the Precious Metals from the Earliest Times to the Present* (New York: Cambridge Encyclopaedia Company, 1901); 2d ed., rev. (1902). This volume was reprinted by Augustus M. Kelly in New York in 1969.

Echevarría, L. Martín, "La leyenda dorada sobre la riqueza de México," *Investigación Económica,* vol. XIV, pp. 231–283.

Esteves, L., *Apuntes para la historia económica del Perú* (Guayaquil, 1882).

Fairnie, D. A., "Commercial Empire of the Atlantic, 1607–1783," *Economic History Review,* vol. XV (1962), pp. 205–218.

Fonseca, Fabián de, and Carlos de Urrutia, *Historia general de la real hacienda* (1791), 6 vols. (Mexico: Impr. por V. G. Torres, 1845–1853).

Furtado, Celso, *Formação económica do Brasil* (Rio de Janeiro: Editora Fundo de Cultura, 1959). An English edition translated by R. W. de Aguiar and E. C. Drysdale was published under the title *The Economic Growth of Brazil: A Survey from Colonial to Modern Times* by the University of California Press at Berkeley in 1965.

Gómez de Cervantes, Gonzalo, *La vida económica y social de Nueva España al finalizar el siglo XVI* (1599), Alberto María Carreño, ed., Biblioteca Histórica Mexicana de Obras Inéditas, no. 19 (Mexico: Antigua Librería Robredo, de J. Porrúa e hijos, 1944).

Hamilton, Earl J., "American Treasure and the Rise of Capitalism, 1500–1700," *Economica,* vol. IX (1929), pp. 338–357.

______, *American Treasure and the Price Revolution in Spain, 1501–1650* (Cambridge, Mass.: Harvard University Press, 1934).

Haring, Clarence H., "The Early Spanish Colonial Exchequer," *American Historical Review,* vol. XXIII (July 1918), pp. 779–796.

______, "Ledgers of the Royal Treasurers in Spanish America in the Sixteenth Century," *Hispanic American Historical Review,* vol. II (1919), pp. 173–187.

______, *Trade and Navigation between Spain and the Indies in the Time of the Hapsburgs,* Harvard Economic Studies, vol. XIX (Cambridge, Mass.: Harvard University Press, 1918).

Helmer, Marie, "La vie économique au XVIe siècle sur le haut plateau andin," *Travaux de l'Institut Français d'Études Andines* (Paris and Lima), vol. III (1951), pp. 115–147.

Levene, Ricardo, *Investigaciones acerca de la historia económica del virreinato del Plata* (Buenos Aires: Casa Editora "Coni," 1927–1928).

Liesegang, Carl, *Deutsche Berg und Huttenleute in Sud und Mittel–Amerika. Beiträge zur Frage des deutschen Einflusses, auf die Entwicklung des Bergbaus in Lateinamerika* (Hamburg, 1949).

López Rosado, Diego G., *Ensayos sobre historia económica de México* Colección Cultura Mexicana, no. 17. 3d ed. (Mexico: Universidad Nacional Autónoma de México, 1965).

———, *Problemas económicos de México,* 2d ed. (Mexico: Instituto de Investigaciones Económicas, Universidad Nacional Autónoma de México, 1966).

Maniau Torquemada, Joaquin, *Historia de la real hacienda de Nueva España* (1794) (Mexico: Alberto M. Carreno, 1914).

Normano, João F., *Brazil, a Study of Economic Types* (Chapel Hill: University of North Carolina Press, 1935).

Rojas, Casto, *Historia financiera de Bolivia* (La Paz: Talleres Gráficas "Marinoni" de A. Chiocchetti, 1916).

Romero, Emilio, *Historia económica y financiera del Perú: Antiguo Perú y Virreynato* (Lima: Imprenta de Torres Aguirre, 1937).

Romero de Terreros y Vinent, Manuel, *Antiguas haciendas de México* (Mexico: Editorial Patria, 1956).

Shafer, Robert J., *The Economic Societies in the Spanish World (1763–1821)* (Syracuse, N.Y.: Syracuse University Press, 1958).

Simonsen, Roberto C., *História econômica do Brasil, 1500–1820,* Biblioteca Pedagógica Brasileira, ser. 5: Brasiliana, 2 vols. (São Paulo: Companhia Editora Nacional, 1937), vols. 100–100A.

Simpson, Lesley B., *The Encomienda in New Spain: The Beginning of Spanish Mexico,* rev. and enlarged ed. (Berkeley: University of California Press, 1950).

Sluiter, Engel, "Francisco López de Cervantes' Historical Sketch of Fiscal Administration in Colonial Peru, 1533–1618," *Hispanic American Historical Review,* vol. XXV (1945), pp. 224–56.

Spix, John B. von and C. F. P. von Martius, *Travels in Brazil in the Years 1817–1820 Undertaken by Command of His Majesty The King of Bavaria,* H. E. Lloyd, trans., 2 vols. (London, 1824). The Portuguese translation by Lúcia Furquim Lahmeyer is entitled *Viagem pelo Brasil,* B. F. Ramiz Galvão e Basílio de Magalhões, rev., Basílio de Magalhões, notes, 4 vols. (Rio de Janeiro, 1938).

Ugarte, César Antonio, *Bosquejo de la historia económica del Perú* (Lima: Imprenta de Cabieses, 1926).

Veitia Linage, José de, *Norte de la contratación de las Indias Occidentales* (1672), 2 vols. in 1 (Seville: I. F. Blas, 1672). This work was reprinted in Buenos Aires in 1945 by the Comisión Argentina de Fomento Interamericano. An English edition translated by Capt. John Stevens was published under the title *The Spanish Rule of Trade to the West Indies: Containing an Account of the Casa de Contratación, or India-House . . .* in London by S. Crouch in 1702. The classic description of Spain's control of trade with her colonies.

Vicens Vives, Jaime, *An Economic History of Spain,* F. M. López-Morillas, trans. (Princeton, N.J.: Princeton University Press, 1969).

———, *Historia social y económica de España y América,* 3 vols. (Barcelona: Editorial Teide, 1957–1959).

Biographies

The following biographies sketch the lives of men who were either directly engaged in mining and metallurgy or influenced some important aspect of the industry. Almost all these men and their works are touched upon in the text.

Aiton, Arthur S., *Antonio de Mendoza, First Viceroy of New Spain* (Durham, N.C.: Duke University Press, 1927).

Alessio Robles, Vito, *El Ilustre Maestro Andrés Manuel del Río* (Mexico, 1937).

Arnade, Charles W., and Kühnel, Josef, "En torno a la personalidad de Tadeo Haenke," *Revista Chilena de Historia y Geografía,* no. 124 (1959), pp. 133–211.

Arnáiz y Freg, Arturo, "D. Fausto de Elhuyar y de Zubice," *Ingeniería* (Mexico), vol. XIV, no. 1 (1940), pp. 1–5.

______, D. Fausto de Elhuyar y Zubice, *Revista de Historia de América,* no. 6 (August 1939), pp. 75–96.

______, *Andrés Manuel del Río* (Mexico: Casino español de México, 1936).

______, Bargalló, Modesto, "Alvaro Alonso Barba: Su vida y su obra científica," *Ciencia* (Mexico), vol. XXVII (1969), pp. 39–42.

Carneiro de Mendonça, Marcos, *O Intendente Câmara* (Rio de Janeiro, 1933). A second edition was published in São Paulo in 1958.

Cox, Patricia, *Don Francisco de Ibarra, capitán de la paz: Apuntes para una biografía* (Culiacán, Mexico: Universidad de Sinaloa, 1965).

Esquivel Obregón, Toribio, *Biografía de Don Francisco Jávier Gamboa* (Mexico: Talleres gráficos Laguna, 1941). This study also appeared in Sociedad Mexicana de Geografía y Estadística, *Boletín,* vol. LIV, nos. 9–10 (1941), pp. 1–233.

Gálvez-Cañero y Alzola, Augusto de, *Apuntes biográficos de D. Fausto de Elhuyar y de Zubice* (Madrid, 1933). This study also appeared in the *Boletín del Instituto*

Geológico y Minero de España (Madrid), vol. LIII (1933).

Gicklhorn, Renée, *Thaddäus Haenkes Reisen und Arbeiten in Südamerika Nach Dokumentar Forschungen in Spanischen Archiven,* Acta Humboldtiana; Series Historica, no. 1 (Wiesbaden, Germany: F. Steiner, 1966).

Hanke, Lewis, *Bartolomé Arzáns de Orsúa y Vela's History of Potosí,* The Colver Lectures in Brown University (Providence, R.I.: Brown University Press, 1965).

———, *Luis Capoche y la Historia de Potosí, 1545–1585,* Pontifical Universidad Católica del Perú, Publicaciones del Instituto Riva-Agüero, no. 28, Separata no. 5, *Cuadernos de Información Bibliográfica* (Lima, 1959).

Kühnel, Josef, *Thaddaeus Haenke: Leben und Leistung eines sudetendeutschen Naturforschers* (Haida, Sudetengau, 1939).

Leonard, Irving A, *Don Carlos de Sigüenza y Góngora* (Berkeley: University of California Press, 1929). A biography of the leading Mexican-born savant of the seventeenth century.

Levillier, Roberto, *Don Francisco de Toledo, supremo organizador del Perú: Su vida, su obra (1515–1582), Colección de publicaciones históricas de la Biblioteca del Congreso Argentino,* 2 vols. (Buenos Aires, 1935).

Liesegang, Carl, *Deutsche Berg- und Hüttenleute in Süd- und Mittelamerika: Beiträge zur Frage des deutschen Einflusses auf die Entwicklung des Bergbau in Latein-amerika,* Hamburger Romantische Studien, B. Ibero-Amerikanische Reihe, vol. XIX (Hamburg: Hansischer Gildenverlag, 1949).

Luanco, Juan R. de, "Los metalurgistas españoles en el Nuevo Mundo," *Crónica Científica* (Barcelona), vols. IX–XI (1886–1888).

Mecham, J. Lloyd, *Francisco de Ibarra and Nueva Vizcaya* (Durham, N.C.: Duke University Press, 1968), reprinted in New York by Greenwood Press in 1968.

Mendoza, Gunnar L., *El Doctor Don Pedro Vicente Cañete y su Historia Física y Política de Potosí* (Sucre: Universidad de San Francisco Xavier, 1954).

Muro, Luis, "Bartolomé de Medina, introductor del beneficio de patio en Nueva España," *Historia Mexicana,* vol. XIII, no. 4 (April–June 1964), pp. 517–531.

Priestley, Herbert I, *José de Gálvez, Visitor-General of New Spain, 1765–1771* (Berkeley: University of California Press, 1916).

Prieto, Carlos, et al., *Andrés Manuel del Río y su obra científica* (Mexico: Cía. Fundidora de Fierro y Acero de Monterrey, 1966).

Probert, Alan, "Bartolomé de Medina: The Patio Process and the Sixteenth Century Silver Crisis," *Journal of the West,* vol. VIII, no. 1 (January 1969), pp. 90–124.

Ramirez, Santiago, *Biografía del Sr. D. Andrés Manuel del Río: Primer Catedratico de Mineralogía del Colegio de Minería* (Mexico: Imprenta del Sagrado Corazón de Jesús, 1891).

Romero de Terreros y Vinent, Manuel, *El Conde de Regla, Creso de la Nueva España,* Vidas Mexicanas no. 9 (Mexico: Ediciones Xochitl, 1943).

———, *Los Condes de Regla: Apuntes Biográficos* (Mexico: Imprenta y Fotografia de M. L. Sanchez, 1909).

Ryden, Stig, *Don Juan José de Elhuyar en Suecia (1781–1782) y el descubrimiento del tungsteno: Apuntes presentados en conmemoración del segundo centenario de su nacimiento al dia 15 de junio de 1754* (Madrid: Insula, 1954).

Shäfer, Ernst, "Johann Tetzel, ein deutscher Bergmann

in Westindien zur Zeit Karl V," *Ibero-Amerikanisches Archiv,* vol. X (1936–1937), pp. 160–170.

Sousa, Otávio Tarquinio de, *José Bonifacio: 1763–1838* (Rio de Janeiro, 1945).

Tamayo y Castillejo, Jorge L., "Don Fausto Elhuyar," *Ingeniería* (Mexico), vol. IX, no. 1 (1935).

Toussaint, Manuel, *Don José de la Borda restituido a España* (Mexico, 1933).

Velásquez de León, Joaquin, "Elogio funebre del Sr. D. Andrés del Río," *El Minero Mexicano,* Feb. 2, 1884.

Vignale, Pedro Juan, "Historiadores y cronistas de la Villa Imperial," *Boletín del Instituto de Investigaciones Historicas* (Buenos Aires), vol. XXVII (1942–1943), pp. 114–130.

Von Hagen, Victor Wolfgang, *South America Called Them; Explorations of the Great Naturalists: La Condamine, Humboldt, Darwin, Spruce* (New York: A. A. Knopf, 1945).

Whitaker, Arthur Preston, "Antonio de Ulloa," *Hispanic American Historical Review,* vol. XV, no. 2 (May 1935), pp. 155–194.

———, "More about Fausto de Elhuyar," *Revista de Historia de América,* no. 10 (December 1940), pp. 125–130.

Yañez, Agustín, *Fray Bartolomé de Las Casas: el Conquistador Conquistado,* Vidas Mexicanas no. 5 (Mexico: Ediciones Xochitl, 1942).

Zimmerman, Arthur Franklin, *Francisco de Toledo, Fifth Viceroy of Peru, 1569–1581* (Caldwell, Idaho: The Caxton Printers, Ltd., 1938).

Law

This section is divided into two parts. The first is a guide to the major law codes and collections, many of which were applied to the New World until superseded by other specific legislation. The second is a collection of commentaries giving either the historical background of mining legislation or discussions of its meaning and application. Several items—Bainbridge, Hoover, Isay, and Van Wagenen—place the field of mining law in general into perspective in relation to political and economic philosophy, property systems, and mining methods. The classic works on mining systems are the studies of Gamboa, Antonil, and Solórzano; excellent modern works are those of Calógeras, Ots Capdequi, and Velarde.

A. Laws and Collections of Laws

Aiton, Arthur S, "First American Mining Code," *Michigan Law Review,* vol. XXIII, no. 2 (1924), pp. 105–114.

______, "Ordenanças hechas por el Sr. Vissorey Don Antonio de Mendoça [sobre las minas de la Nueva España: Año de M.D.L.]," *Revista de Historia de América,* no. 14 (June 1942), pp. 73–95.

Cañete y Domínguez, Pedro Vincente, *Ordenanzas de las minas del Perú y demas provincias del Río de la Plata* (1794).

Chism, Richard E., "Synopsis of the Mining Laws of Mexico," American Institute of Mining Engineers, *Transactions* (New York, 1902).

The Civil Law, Including the Twelve Tables, the Institutes of Gaius, the Rules of Ulpin, the Opinions of Paulus, the Enactments of Justinian and the Constitutions of Leo, S. P. Scott, trans. and ed. (Cincinnati: The Central Trust Co., 1932).

Encinas, Diego de, *Provisiones, cédulas, capítulos de ordenanzas, instrucciones, y cartas . . . tocantes al buen gobierno de las Indias, y la administración de la justicia en ellas; sacado todo ello de los libros del dicho Consejo,* 4 vols. (Madrid, 1596).

Halleck, Henry W., *A Collection of Mining Laws of Spain and Mexico* (San Francisco, 1859).

Martínez Alcubilla, Marcelo, ed., *Códigos antiguos de España: Colección completa* (Madrid, 1885).

———, *Códigos españoles* (Madrid, 1872).

Murillo Velarde, Pedro, ed., *Cursus juris canonici, hispani, et indici* (Madrid, 1791).

Novisima Recopilación de las leyes de España 4 vols. (Madrid, 1805–29).

Nueva Recopilación de las leyes de España (Madrid, various editions and dates from 1537 to 1777).

Ordenanzas de minería [1783] y colección de las órdenes y decretos de esta materia, (Paris: Ed. Bouret, 1851). A later edition, issued just before the repeal of the ordenanzas in 1884, is Nueva Edición Dispuesta por C.N. (Paris and Mexico: Librería Ch. Bouret, 1881).

Pereira Salas, Eugenio, "Las ordenanzas de minas del gobernador de Chile, don Francisco de Villagra," *Revista de Historia de América,* no. 23 (December 1951), pp. 207–225.

Peres, Damião, *Antecedentes da legislacão concernente ao ouro do Brasil nos seculos XVI e XVIII* (Lisbon: Academia Portuguêsa de História, 1956).

Polo Ondegardo, Juan, "Ordenanzas de las minas de Guamanga" (Mar. 25, 1562), *Colección de documentos inéditos relativos al descubrimiento . . . de América y Oceanía,* vol. VIII (Madrid, 1867), pp. 449–462.

Puga, Vasco de, *Provisiões cédulas instrucciones de Su Magistad . . . ñaças d' difũtos y audiẽcia, pa la buena expediciõ de los negocios, y administraciõ d' justicia: Y gouernaciõ d' sta Nueva España: Y pa el buẽ tratamiẽto y osevaciõ d' los Yndios, dende en el año 1525. hasta este presente de .63* (Mexico: Casa de Pedro Ochante, 1563). This collection of early Spanish law was reprinted by

Ediciones Cultura Hispanica in a facsimile edition in its series *Colección de Incunables Americanos Siglo XVI* as vol. III in Madrid in 1945.

Real Orden de 8 de diciembre de 1785, y declaraciones en su cumplimiento hechas para adaptar la Ordenanza de Minería de Nueva-España a el virreinato de Lima (Lima, 1786).

Recopilación de las leyes de los Reynos de las Indias (Madrid, 1681); 5th and final edition, 4 vols. (1841).

Rockwell, John A., *A Compilation of Spanish and Mexican Law in Relation to Mines and Titles to Real Estate* (New York, 1851).

Las Siete Partidas, Samuel P. Scott, trans. Published in collaboration with the Comparative Law Bureau of the American Bar Association (New York: Commerce Clearing House, 1931).

Toledo, Francisco de (Viceroy), "Ordenanzas del . . . para los oficiales reales de Guaymanga y Caja de Guancavelica" (1582). Luis Torris de Mendoza, ed. *Colección de documentos inéditos relativos al descubrimiento . . . de América y Oceanía,* vol. VIII (Madrid, 1867), pp. 462–484. The *Ordenanzas del Perú is reprinted in Emilio Tagle Rodriguez, Legislación de minas* (Santiago de Chile: Imprenta Chile, 1918).

The Visigothic Code, S. P. Scott, trans. and ed. (Boston: The Boston Book Co., 1910).

White, Joseph W., *A New Collection of Laws, Charters and Local Ordinances of the Governments of Great Britain, France, and Spain, Relative to the Concessions of Land in their Respective Colonies* (Philadelphia, 1839).

B. Commentaries on Mining Law

Bainbridge, William, *A Treatise on the Law of Mines and Minerals* (Philadelphia: J. Campbell, 1871).

Barrios, B., "The Historical Origin of the Spanish Legislation in Matters of Mining . . . ," Association for

International Law, 28th Conference, Madrid, 1913, *Compte Rendu* (Madrid, 1914).

Bonifaz, Miguel, "Minería y legislación colonial," *Revista de Estudios Juridicos, Políticos y Sociales* (Sucre), vol. X (1950), pp. 195–224.

Calógeras, João Pandía, *As minas do Brasil e a sua legislação,* 3 vols. (Rio de Janeiro, 1904–1905). This study can be used in conjunction with Eschwege's chapters on mining law in Brazil in *Pluto Brasiliensis.*

Gamboa, Francisco Xavier de, *Comentarios a las ordenanzas de minas dedicados al católico rey, maestro Señor Don Carlos III* (Madrid: J. Ibarra, 1761). An English translation was made by R. Heathfield under the English title *Commentaries on the Mining Ordinances of Spain* in two volumes. It was published in London by Longman, Rees and Rowe in 1830.

Hoover, Herbert C. and Hoover, L. H., "Notes on the Development of Mining Law," *Engineering and Mining Journal,* vol. XCIV (Nov. 2, 1912), pp. 823–825.

Isay, Rudolf, "Mining Law," *Encyclopaedia of the Social Sciences,* vol. X (New York: The Macmillan Co., 1933), pp. 513–517.

Lassaga, Juan Lucas de, and Velásquez de León, Joaquin, *Representación que a nombre de la minería de esta Nueva España, hacen al rey nuestro señor los citados señores, apoderados de ella, Regidor, el primero, de la Novilísima ciudad de México y Juez Contador de Mensores y Albaceazgos; y abogado el segundo de su Real Audiencia y Catedrático que has sido de matemáticas de su Real Universidad* (Mexico, 1774).

Martínez Baca, Eduardo, *Reseña histórica de la legislación minera en México* (Mexico: oficina Tipográfica de la Secretaría de Fomento, 1901). This study was translated as "Historical Sketch of Mining Legislation in Mexico" and printed in the American

Institute of Mining Engineers *Transactions,* vol. XXXII (1902), pp. 520–565.

Martiré, Eduardo, "El derecho minero en la época de la Independencia (1810–1820): Contribucion para su estudio," *Revista del Instituto de Historia del Derecho Ricardo Levene*, no. 17 (Buenos Aires, 1966) pp. 41–88.

———, *Panorama de la legislación minera argentina en el período hispánico* (Buenos Aires, 1968).

Ots Capdequí, José María, *Estudios de historia del derecho español en las Indias* (Bogotá, 1940).

Peña, Manuel de la, *El dominio directo del soberano en las minas de México y génesis de la legislación petrolera mexicana,* 2 vols. (Mexico, 1928).

Solórzano Pereira, Juan de, *Politica indiana: Sacada en lengua castellana de los dos tornos del derecho, i govierno municipal de las Indias Occidentales que mas copiosamente escribo en la latina [por . . .]* (Madrid: D. Díaz de la Carrera, 1648). This study was first published in Latin in two volumes in Madrid in 1629–1639 under the title *De Indiarum ivre . . . .* There is a reprint edition by Compañía Ibero-American de Publicaciones, Madrid and Buenos Aires: 1945.

Van Wagenen, Theodore Francis, *International Mining Law* (New York: McGraw-Hill Book Company, 1918).

Vance, John T., *The Background of Hispanic-American Law: Legal Sources and Juridical Literature of Spain* (New York: Central Book Co., 1943).

———, and Clagett, Helen L., *A Guide to the Law and Legal Literature of Mexico* Latin American Series, no. 6 (Washington, D.C.: Library of Congress, The Law Library, 1945).

Velarde, Carlos E., *Historia del derecho de minería hispano-americano y estado de la legislación de minas petróleo . . .*

(Buenos Aires: Talleres gráficos argentinos de L. J. Rosso y Cía., 1919).

———, *La legislación minera en las repúblicas hispano-americanas: La propiedad minera; su orígen, caracteres y condición resolutoria* (Buenos Aires: Compañía Sud-Americana de Billetes de Banco, 1916).

Miscellaneous Studies

Alamán, Lucas, *Disertaciones sobre la historia de la República mexicana, desde la época de la conquista que los Españoles hicieron, a fines del siglo XV y principios del XVI, de las minas y continente americano, hasta la independencia,* 3 vols. (Mexico: Imprenta de J. M. Lara, 1844–1849).

Amador, Elias, *Bosquejo histórico de Zacatecas,* (Zacatecas, Mexico 1892). This history of Zacatecas is useful for studying the growth of the celebrated mining town.

Armando, Nicolau, *Valenciana* (Mexico: Dirección de Monumentos Coloniales, Instituto Nacional de Antropología e Historia, 1961).

Bayle, Constantino, S. J., *El Dorado fantasma* (Madrid: Editorial Razón y Fé, 1930). A second edition was issued in Madrid by the Consejo de Hispanidad in 1943.

Boxer, Charles R. *The Golden Age of Brazil, 1695–1750: Growing Pains of a Colonial Society* (Berkeley: University of California Press, 1962). Excellent treatment of mining for gold and diamonds in eighteenth century Brazil.

———, *Salvador de Sá and the Struggle for Brazil and Angola, 1602–1686* (London: University of London, Athlone Press, 1952). Contains a chapter on "The Road to Potosí."

Cerqueira Falção, Edgard de, *Relíquias da terra de ouro* (São Paulo, 1946).

Cortés, Antonio, *Valenciana: Guanajuato, Mexico* (Mexico: Secretaría de Educación Pública, 1933).

Cortesão, Jaime, *Rapôso Tavares e a formação territorial do Brasil* (Rio de Janeiro, 1958).

______, "A maior bandeira do maior bandeirante," *Revista de História,* vol. XXII, no. 45 (January–March 1961), pp. 3–27.

Diffie, Bailey W., *Latin American Civilization: Colonial Period* (Harrisburg, Pa.: Stackpole Sons, 1945); reprinted in New York by Octagon Books in 1967. Devotes several chapters to mining.

Ellis, Alfredo, Jr., *O ouro e a Paulistânia* (São Paulo, 1953).

Fernández, Justino, *El Palacio de Minería* (Mexico City: Instituto de Investigaciones Estéticas, Universidad Nacional Autónoma de México, 1951).

Larraz López, José, *La época del mercantilismo en Castilla, 1500–1700,* 2d ed. (Madrid: Ediciones Atlas, 1943). Discusses the effect of silver and gold in Castile.

Lima, Augusto da, Jr., *A Capitania das Minas Gerais,* 2d ed. (Rio de Janeiro, 1943).

______, *Vila Rica do ouro preto; Sintese histórica e descritiva* (Belo Horizonte, 1957).

Maltby, William S., *The Black Legend in England* (Durham, N.C.: Duke University Press, 1971).

Mesa, Jose de and Teresa Gisbert de Mesa, "Un pintor colonial boliviano: Melchor Pérez Holguín," *Arte de America y Filipinas,* vol. II, no. 4 (Seville, 1952), pp. 149–216.

Morse, Richard M., *The Bandeirantes: The Historical Role of the Brazilian Pathfinders* (New York: A. A. Knopf, 1965). The best study in English about the bandeirantes. Reproduces a number of contemporary documents.

Stubbe, Carlos F., *Vocabulario minero antiguo: Compilación de términos antiguos usados por los mineros y metalurgistas de la América Ibérica* (Buenos Aires: Tall. Graf. M. Violetto, 1945).

Taunay, Alfonso de Escragnolle, *História geral das bandeiras Paulistas: Escripta á vista de avultada documentação inedita dos archivos brasileiros, hespanhoẽs e portuguezes,* 8 vols. (São Paulo: H. L. Canton, 1927–1950). The basic work on the subject because of the copious citations from original documents. An abridged version in two volumes was published in São Paulo in 1954 under the title *Historia das bandeiras Paulistas.*

Torres, João Camillo de Oliveira, *História de Minas Gerais,* 5 vols. (Belo Horizonte, 1961).

Toussaint, Manuel, *Guia illustrada de Tasco* (Mexico City, 1935). Includes an English translation.

——, *Tasco: Su historia, sus monumentos* (Mexico City, 1931).

——, "La Casa de Moneda de Potosí," *Boletín de la Sociedad Geográfica* (Potosí, 1940).

Wolff, Inge, "Zur Geschichte der Ausländer in spanischen Amerika. Die Stellung des Extranjero in der Stadt Potosí vom 16. bis 17. Jahrhundert" in Otto Brunner, ed., *Europa und Ubersee. Festschrift Für Egmont Zechlin* (Hamburg: Verlag Hans Bredow–Institut, 1961), pp. 78–108.

Index